EUGENICS SOCIETY SYMPOSIA

Volume 4

GENETIC AND
ENVIRONMENTAL
INFLUENCES
ON BEHAVIOUR

EUGENICS SOCIETY SYMPOSIA

GENETIC AND ENVIRONMENTAL INFLUENCES ON BEHAVIOUR

A Symposium held by the Eugenics Society in September 1967

Edited by

J. M. THODAY

A. S. PARKES

SPRINGER SCIENCE+BUSINESS MEDIA, LLC

Library of Congress Catalog Card Number: 68–54003

ISBN 978-1-4684-8651-3   ISBN 978-1-4684-8649-0 (eBook)
DOI 10.1007/978-1-4684-8649-0

First Published 1968
© 1968 Springer Science+Business Media New York
Originally published by Plenum Press in 1968
Softcover reprint of the hardcover 1st edition 1968

EDITORS' FOREWORD

THIS volume records the proceedings of the fourth Symposium organized by the Eugenics Society. Like its predecessor, it deals with the interaction of genetic and environmental factors, in this case their interaction to produce behavioural patterns, especially in man. Possibly in no field is the interaction of genetic and environmental factors more evident. Behaviour is determined by the impact of environment on present personality resulting from inborn forces modified by past environment. Under certain conditions, such as chromosome abnormality, genetical factors dominate the situation. Under others, such as maternal deprivation, the early social environment may be the major factor, especially in sexual behaviour. In the event, the Symposium produced many outstanding contributions, and the meeting may be said to have succeeded admirably in the avowed object of the series—tilling the common ground between biology and the social sciences. The Society's thanks are thus again due, as on previous occasions, to all the participants. It is unfortunate that it has not been possible to include a record of the Chi-Chi film. Professor Kalmus, however, has kindly provided a script of his commentary on the film and tape-recording which formed a fascinating finale to the Symposium.

The success of the four Symposia so far held in this series, and the wide appeal of the published proceedings, has encouraged the Council of the Eugenics Society, through the medium of the newly established Galton Foundation, to launch a *Journal of Biosocial Science*, the first number of which will appear in January 1969. Lord Platt, in his general introduction (p. ix), has outlined the projected scope of the journal.

The Editors-in-Chief will be Professor E. Grebenik and Dr. C. O. Carter, and they will be supported by a strong Editorial Board. It is anticipated that the new journal will give focus and perspective to this vitally important but at present rather ill-defined field. Its appearance will not add to the current publication explosion, because *The Eugenics*

Review, now in its sixtieth year and having fulfilled its object, will be discontinued at the end of 1968. The Jubilee volume is being devoted to historical matter and cumulative indices.

As the two members of Council mainly responsible for arranging the programme, we have accepted formal responsibility for the editing of this volume, but—as with previous ones—the day-to-day work has fallen on Mrs. K. Hodson, now retired from the Society's service with the good wishes of all, but still working privately on the Symposium volumes. Our best thanks to her.

J. M. THODAY
A. S. PARKES

CONTENTS

viii CONTENTS

GENERAL INTRODUCTION

LORD PLATT

Clinical Research Board, Medical Research Council, London

THE more that science grows and knowledge accumulates, the more it is bound to fragment into various parts, because specialization becomes so necessary to those who, by their researches, are adding still further to knowledge; at the same time, the more necessary it becomes for links to be forged and communications to be wide open between the various branches of science. The Eugenics Society has felt that it has an almost unique opportunity for helping to make real those links between the various branches of science which are concerned with man, and particularly man's future.

The aims of the Society have been stated as " To study inborn human qualities, to formulate and support policies for improving these qualities and enabling them to develop to their full potential in the individual; to foster a responsible attitude to parenthood, promote relevant research, and facilitate communication between those interested." It is the current policy of this Society to bring together the many different disciplines—genetics, biology, sociology, medicine, demography, economics—which bear on these aims.

During the last year or so, the Eugenics Society has had two new ideas on how it can further these aims : one is to start a new Foundation—the Galton Foundation. It is just sixty years since this Society was started, so that it seems an appropriate time to embark on a new venture. The Foundation will be sponsored by the Eugenics Society but will be an independent organization, which we hope will attract more funds for research and for the promotion of our objects.

At the same time, and through the medium of the Galton Foundation, we are launching a new internationally orientated journal—though we know it is not always popular to start a new journal nowadays—and preliminary notices of the *Journal of Biosocial Science* have been issued. Its purpose will be to focus attention on, and promote communication within, the fields of human biology and sociology. We believe that there is no other journal at the present time which fulfils these particular needs. We hope to publish original papers, reviews, lectures, proceedings and major book reviews dealing with social aspects of human biology, including reproduction and fertility-control, gerontology, nutrition, ecology, genetics and applied psychology and ethology; with biological aspects of the social sciences, including sociology and social studies, social anthropology and ethnology, economics, education and criminology; and with biological, sociological, economic and historical aspects of demography.

In relation to these various functions of the Eugenics Society, I would like to say how much the Society owes to its Honorary Secretary, Professor (now Sir Alan) Parkes.

As regards the present Symposium, the subject this year is *Genetic and Environmental Influences on Behaviour*. Our previous Symposia have been very popular and very successful. This is proved by the way that the published papers have been in very considerable demand and the fact that all the tickets for this Symposium were taken up soon after the programme was announced.

The theme of this Symposium arose out of discussions between several members of the Council of the Eugenics Society, but the idea of it came particularly from Professor Thoday, whose subject is genetics, and who is Chairman of Session III. He felt that sufficient work was now being done on the genetics of behaviour to make this a good subject for a symposium. But of course the material to be presented goes beyond the field of genetics, as Professor Thoday himself suggested that it should; and the contents list of this volume shows the various ways in which contributors discuss genetic and environmental influences on behaviour.

COMPARATIVE STUDIES

Chairman: LORD PLATT

DICHOTOMIES IN THE STUDY
OF DEVELOPMENT

R. A. HINDE

Sub-Department of Animal Behaviour, Madingley, University of Cambridge

THE role of first speaker in this Symposium is, I suppose, to provide a conceptual backdrop to what follows. It seems a pity to delay discussion of data by thrashing over yet again the dreary old nature-nuture problem; and yet it is one which still gives rise to bloody pates amongst students of behaviour. Since the blows are administered by men of good faith who believe in their own views, it is worth while to see how the disagreements arise.

Mostly, I believe, they stem from the use, or rather the misuse, of dichotomies. Inevitably, and properly, those who work on the development of behaviour use classification as a tool, and it has often taken the form of dividing the subject matter into categories of 'instinctive' versus 'intelligent', or 'innate' versus 'learnt'.

Now there is one immediate difficulty with all such dichotomies: they can be taken to imply that the factors influencing the development of behaviour are of two types only—genetic, and those associated with learning. This is of course fallacious, for environmental factors can influence the course of development in many ways, some of which do not come within any generally accepted meaning of 'learning'. For example, a given genetic strain of *Drosophila melanogaster* may be capable of full normal flight, or of erratic flight, or may be incapable of flying at all, depending on the temperature during development;[3] and the amount of handling a rat receives in infancy may affect the amount it moves about and its tendency to defecate when placed in an open field in adulthood.[10] In neither of these cases can the effects of the environment usefully be described simply as 'learning': in the first, the environmental influence is (probably) not even exerted through the sense organs;

and in the second, the effect is mediated at least in part through a change in the functional properties of the endocrine system. Although learning can be defined so broadly that it includes all environmentally-induced modifications to behaviour mechanisms, such a practice is not only contrary to common usage, but could make the concept of learning useless as an analytical tool. Lorenz,[12] following Thorpe,[16] comes near to doing this, for he defines learning to include " all adaptive modifications of behaviour ". In so doing he treats learning as a loose concept, hard at the core but blurred at the edges: and in other contexts he is driven to distinguish the core as ' true learning ', and to regard it as " distinct from other processes of adaptive modification ". Learning is indeed a loose concept, but so long as a precise delimitation in behavioural terms remains impossible, it seems desirable to adhere to common usage. However, it is clear that we are not going to be bothered with this difficulty here; for the organizers have carefully entitled this Symposium *genetic and environmental influences on behaviour*.

A more fundamental difficulty with dichotomies, and one which I believe to lie at the root of much of the trouble, arises from the application by different workers of the *same* dichotomy to *different* things. Science depends on classification but, in any particular context, it is necessary to agree not only about the system of classification, but (is it necessary to add ?) also about what it is that is being classified.

Traditionally, dichotomies of the types under discussion were applied to behaviour itself. A bird's nest-building behaviour was described as ' instinctive ', with its delevopment supposedly uninfluenced by environmental factors, and contrasted with ' intelligent ' behaviour, like building houses, which requires learning. This sort of distinction was perhaps useful at a superficial level of analysis, but it did not bear close inspection.[1, 4, 5, 8, 9, 13, 14, 18] ' Instinctive ' or ' innate ' behaviour was usually defined solely in negative terms: a response was said to be innate or instinctive just so long as no environmental influence on its development had been identified. Definitions in negative terms have limited usefulness; for the categories so defined could themselves have almost unlimited heterogeneity. Furthermore, such definitions ran into the difficulty of proving that all environmental factors were without influence. Often the classification depended on a ' deprivation experiment '—animals reared in an environment lacking some factor believed to be important were compared with animals growing up normally. But no such deprivation experiment, or series of deprivation experiments, could exclude *all* environmental influences. In any case, it is a truism that all behaviour depends on both nature and nurture. Each unit or character of behaviour, just like each somatic character and indeed the body itself, can develop only within a certain range of environmental conditions. Of course, while some types or aspects of behaviour develop only under very specific environmental conditions, others appear under any conditions in which life itself is possible. For this reason, it is sometimes useful to characterize items of behaviour according to their stability or lability under environmental influences: this, however, is not to erect a dichotomy, but to recognize a continuity.

A second type or dichotomy is applied not to the behaviour itself, but to the processes involved in its development. Thus it is often useful to distinguish the consequences of ' maturation ' (that is, tissue growth and differentiation) from those of ' experience ' (that is, influences of the extraorganismic environment). But this distinction is useful only so long as we do not forget that it is merely a convenient abstraction[15] which depends on the extent of the universe being considered. Processes of maturation occur in, and depend on, the intraorganismic environment; they are not influenced by extraorganismic factors, largely because the environment relevant to them is maintained constant by homeostatic mechanisms. Furthermore, the consequences of maturation often vary with the experience of the individual: the eye may form through processes of maturation, but not be functional until exposed to light.[6]

Lorenz[11, 12] attempts to overcome some of the difficulties inherent in a dichotomy of behaviour by using a third type of dichotomy—one between 'sources of information '. Emphasizing that behaviour is by and large adaptive in a biological sense—that is, it increases the individual's chances of survival and reproduction—he argues that adaptedness involves acquisition of information about the environment, and this occurs in only two ways. First, the species achieves adaptedness through mutation and natural selection; and second, the individual acquires information by interaction with its surroundings. In the individual the information necessary for the development of adapted behaviour thus comes from only two sources—the genes, moulded by the adaptive processes of evolution, and the experience of that individual.

Lorenz thus uses a dichotomy of ' sources of information ' —' the innate ', referring to acquisition by the species in evolution, and ' the learned ', referring to acquisition by the individual in commerce with its environment. As we shall see, some disagreements have arisen partly because Lorenz failed to define sufficiently rigidly what he means by ' the innate '. His statement that " the innate [is] . . . what must be in existence before all individual learning in order to make learning possible . . ." makes it difficult to grasp what sort of a concept ' the innate ' is. Perhaps what he means by ' innate information ' is best clarified by example. He writes:

> All we have to do is to rear an animal, as perfectly as we can, under circumstances that withhold the particular information which we want to investigate. . . . If our baby *Apistogrammas* [a black and yellow Cichlid fish—R. A. H.], who have never seen an adult female of their species, selectively respond to a certain black and yellow pattern by following it and staying with it (as they would with their real mother) or if our stickleback [i.e. one reared in isolation—R. A. H.] responds to an object which is red below with the highly specific motor patterns used in fighting a rival, we are justified in asserting that the information which these fishes possess concerning these two objects is fully innate.

Lorenz claims that his definition of ' the innate ' is " truly operational " and points to the use of " deprivation experiments ". If a male stickleback is reared in isolation from other sticklebacks (" . . . if we withhold from our stickleback all information on the colours of its rival "), and if it subsequently responds differentially and appropriately to the species-characteristic colour pattern, then " . . . we can conclude that the information about the colour pattern has been relayed by the genome ".

Lorenz is thus concerned primarily with the adaptedness of behaviour, and with the two sources of "information" whereby this is achieved. For the most part he is avowedly not interested in the interplay between organism and environment which is the essence of development. He writes:

What we want to elucidate are the amazing facts of adaptedness. . . . We need not bother about the innumerable factors which may cause ' differences ' in behavior, as long as we are quite sure that they cannot possibly relay to the organism that particular information which we want to investigate.

In that Lorenz is not concerned with the detailed ontogeny of behaviour, but only with the ' sources of information ' for adaptedness, his interests are quite different from those of more analytically-minded students of behaviour, whose aim is to tease out the interacting factors involved in development. It is especially important to remember this, for Lorenz himself sometimes forgets that the problem in which he is interested differs from that of most students of behavioural development, and often slides from one type of dichotomy to another. For instance, while in some contexts he agrees that a dichotomy of behaviour is not useful, in others he rejects the criticisms which have been made of such a dichotomy, with arguments which are valid as support for a dichotomy only of the type he is advancing.

Lorenz's approach has an immediate appeal to those who are concerned with the behaviour of animals in nature. It is clearly pertinent in the innumerable cases in which an animal's response to a situation must or can be appropriate in the absence of experience of that situation. Furthermore, his dichotomy throws into relief an important point which is too easily overlooked in studies of development. He points out that the chance that any modification produced by the environment will be adaptive to that environment is infinitesimally small. Thus when we find that an environmental change does produce an adaptive response, as when the thickness of fur increases in a cold environment, then the response is due not only to the environmental change, but also to the specialized nature of the organism's response to that change. In the same way, since behaviour is on the whole adaptive (in a biological sense), some

B

explanation is required for the fact that the environmental factors which operate in the development of that behaviour act to produce results favourable to the organism, and not deleterious ones. The demonstration that the development of a given adaptive pattern of behaviour depends on specific environmental factors can never be the whole story: we must also ask why the developing organism responds to these factors in an adaptive way, and not in some other way. In particular, we must ask why an animal tends to learn what is good for it. Lorenz reviews evidence supporting the view that learning abilities are not generalized and diffuse, but limited to certain sections or aspects of behaviour; that this limitation is in fact adaptive; and that the limitation is ultimately genetically determined. He also argues that, if the development of pecking in chicks does in fact depend on learning before hatching, as Schneirla[14] and Lehrman[8] have argued on the basis of Kuo's[7] data, it remains to be shown why some species learn to peck and others do not; in his view, all learning is performed by mechanisms which contain " phylogenetically acquired information ". It would not be pertinent here to pick minor quarrels with this formulation, though it is worth noting that Lorenz generalizes from a knowledge of the reinforcing effectiveness for many species of food, water and sex, to imply that all situations which can be shown to be reinforcing do in fact produce adaptive consequences—a view which requires a sharpening of the concept of ' adaptive '. His main point here is, however, a valid one.

But here Lorenz runs into danger from his own broad definition of learning. His evidence that some aspects of behaviour are not influenced by learning (in a narrow sense) in no way shows that some aspects of behaviour are independent of environmental influences in the sense that they are infinitely stable. It is here that he slides from a dichotomy of sources of information leading to adaptedness to the old dichotomy of behaviour.

Lorenz applies his dichotomy especially to cases in which appropriate response to a stimulus does or does not depend on previous experience of that stimulus, or the patterning of a movement does or does not depend on practice. Thus the male stickleback reared in isolation has no chance to

COMPARATIVE STUDIES 9

learn what his rival looks like, yet somehow responds appropriately. Understanding may not be advanced greatly by ascribing the appropriateness of the response to ' innate information ', but the nature of the implication is clear. Similarly, if a chaffinch can sing the species-characteristic song only after hearing other chaffinches singing, we can describe this by saying that ' environmental information ' is necessary. In other cases, however, the dichotomy seems less appropriate. Consider a rodent which, after being reared by its mother, responds to a strange situation in a manner and to an extent which are presumably adaptive. We know that the response can be influenced by the amount of disturbance experienced during rearing. But can we describe the amount of disturbance the infant gets in the nest as providing information about the nature of the potentially dangerous situations it will experience in adulthood ? Is it helpful to speak of environmental information furnished by a situation quite different from that in which the response is exhibited ?

Another difficulty with Lorenz's dichotomy seems to result from the beauty of his own metaphors. By picking examples, like that of the male stickleback, in which the role of learning is limited, Lorenz manages to imply that some behaviour depends on a ' blue-print ' contained in the genome which requires only ' building stones ' for its realization. But it is worth while to examine this colourful picture more closely. Lorenz implies that these ' building stones ' are the conditions necessary for normal growth. If ' learning ' is used as referring to all adaptive modifications produced by environmental influences, then what are these ' building stones ' but the substance of learning ? By that definition, learning enters into the development of all such behaviour. If ' learning ' is used in a narrower sense, then these ' building stones ' refer to other environmental influences on behaviour. In either case, if the nature of the resultant behaviour depends at all on the nature of these ' building stones ', it cannot be said to be independent of ' environmental information '.

Furthermore, one must not be blinded by the superficial gloss which accompanies the word ' information '. If it is to be used in a precise way, it must, at least in theory, be possible to measure the amount of information contributed by the genome and that contributed by the environment in the development of any unit of behaviour.[17] Now, to measure the information required to describe a given state or a given pattern of behaviour, we must know the number of states or patterns of behaviour possible. Calculations of the information content of an egg start from knowledge of the proportions of the various atoms present, and must depend on assumptions about those which *might* have been incorporated. To assess the information content of the environment, we must make assumptions about the range of environmental states possible. The difficulty is even greater with behaviour: to specify a chaffinch's song, say, certain parameters must be defined, but who can assess the number of possible parameters which it might have had? And how can we equate the assumptions made in assessing the information content of the pattern of behaviour with those made in assessing that in the zygote or the environment? And if behaviour depends on interaction between organism and environment, is it meaningful to ask how many bits of information depend on genetic factors and how many on environmental? Clearly, Lorenz is using ' information ' in a very general and non-technical way.

Before leaving Lorenz's dichotomy, mention must be made of a difficulty to which it gives rise in the interpretation of deprivation experiments. Lorenz believes that such experiments can justify only assertions about what is *not* learnt. If a stickleback reared in isolation responds appropriately to a rival, " . . . we then know without any further experiment that information concerning the rival's colours has been relayed by the genome ". If, however, the subject fails to respond, Lorenz argues that we cannot assert that the response normally depends on learning, because deprivation may have involved the withholding of a ' building stone ' necessary for the realization of the ' blue-print '. This brings us back to the difficulty of the nature of these building stones, which we have already discussed.

Many would in fact argue exactly the opposite: the so-called deprivation experiment can only tell us that an environmental influence *is* important. If there is a difference in behaviour between animals brought up with, and those brought up without, a given environmental factor, then, other things being equal, that factor has an influence: the nature of that influence and whether exerted through ' learning ', however defined, is a further question. If there is no difference between the two groups, we can assert only that our experiment has revealed no influence of the environmental factor, and not that that factor has no influence under all circumstances.

This brings us to the fourth type of dichotomy which can be applied in the study of the development of behaviour—a dichotomy according to the source of differences. The so-called deprivation experiment involves, in fact, comparing the behaviour of animals brought up with, and without, a given environmental factor. If all other factors (including the genetic constitution of the individuals tested) are controlled, then a difference in behaviour can be ascribed to the difference in the environment between the two groups. If no difference appears, then our experiment reveals no influence of the factor in question, though of course other environmental factors may be important.

The converse experiment is, of course, also possible. If we subject two individuals or groups of individuals which differ genetically to identical rearing conditions, then any differences in behaviour which appear must ultimately be due to the genetic differences.

By such experiments it is possible to associate differences in behaviour between organisms with differences in nature or nurture between them. This is obviously not the same as ascribing particular pieces of behaviour to nature or nurture, nor is it the same as attempting to divide the ' information ' content of each pattern between two sources. It is rather a distinction between differences according to their source, and it is a distinction which is, in theory at any rate, readily made operational.

There are, of course, practical difficulties. It is rarely possible to obtain two individuals which are identical genetically, or which have been reared in identical environments; but this difficulty is easy enough to overcome by using groups of individuals. More serious, perhaps, are the dangers of over-interpretation. The finding that a difference in a particular character between two organisms is due ultimately to a genetic difference between them is no indication whatever that environmental factors played no part in its development. Indeed, the labelling of a difference as genetic or environmental tells us nothing of the ontogeny of the character: all environmental differences operate by permitting or restricting expression of a genetic potentiality, while a genetic difference may be expressed as the presence or absence of response to an environmental factor. The problem becomes particularly acute in species where the young are dependent on maternal care. Adult behavioural characteristics may then depend on the nature of mother-infant interaction in infancy. This in turn depends on both the genetic constitution and the previous environmental history of both mother and infant.[2]

Thus, if we hold the aim of ontogenetic studies to be the unravelling of the pattern of changes occurring in time, data as to the source of a behaviour difference are only a first step. It is necessary to ask further how the genetic or environmental factors in question produce the difference in behaviour. Nevertheless the examination of successive differences, produced by successive types of experimental interference, is the only way of breaking open the ontogenetic interaction.

We have considered four types of dichotomy used in studies of behavioural ontogeny, each of which has value in certain contexts. Dichotomies of behaviour and dichotomies of processes, however, tend to be useful only in the initial stages of analysis, while Lorenz's dichotomy of sources of information, perhaps useful to the evolutionary biologist, is quite inadequate for the detailed study of ontogenetic processes. Thus the dichotomy with most promise as an analytical tool is the dichotomy of sources of differences. This, indeed, is the way in which the title of this Symposium must be applied, because the relevance of a factor to the development of behaviour can be established only by a difference resulting from its presence or absence. In any case, much confusion can be avoided if we attempt to be clear as to the sort of dichotomy we are using.

Acknowledgment

I am grateful to W. H. Thorpe and J. G. Stevenson for their comments on the manuscript.

REFERENCES

1. BEACH, F. A. 1955. The descent of instinct. *Psychol. Rev.* **62**, 401.
2. DENENBERG, V. H. and WHIMBEY, A. E. 1963. Behaviour of adult rats is modified by the experiences their mothers had as infants. *Science* **142**, 1192.
3. HARNLY, M. H. 1941. Flight capacity in relation to phenotypic and genotypic variations in the wings of *Drosophila melanogaster*. *J. exp. Zool.* **88**, 263.
4. HEBB, D. O. 1953. Heredity and environment in mammalian behaviour. *Br. J. Anim. Behav.* **1**, 43.
5. JENSEN, D. D. 1961. Operationism and the question ' Is this behavior learned or innate? ' *Behaviour* **17**, 1.
6. KNOLL, M. 1953. Über das Tages- und Dämmerungssehen des Grasfrosches (*Rana temporaria L.*) nach Aufzucht in veränderten Lichtbedingungen. *Z. vergl. Physiol.* **35**, 42.
7. KUO, Z-Y. 1932. Ontogeny of embryonic behavior in Aves : IV. The influence of embryonic movements upon the behavior after hatching. *J. comp. Psychol.* **14**, 109.
8. LEHRMAN, D. S. 1953. A critique of Konrad Lorenz's theory of instinctive behaviour. *Q. Rev. Biol.* **28**, 337.
9. LEHRMAN, D. S. In *Development and Evolution of Behaviour*. Vol. I. Eds. D. S. Lehrman, J. S. Rosenblatt and E. Tobach. Freeman. (In press.)
10. LEVINE, S. 1962. The effects of infantile experience on adult behavior. In *Experimental Foundations of Clinical Psychology*. Ed. A. J. Bachrach. New York. Basic Books.
11. LORENZ, K. 1961. Phylogenetische Anpassung und adaptive Modifikation des Verhaltens. *Z. Tierpsychol.* **18**, 139.
12. LORENZ, K. 1965. *Evolution and Modification of Behavior*. University of Chicago Press.
13. SCHNEIRLA, T. C. 1951. A consideration of some problems in the ontogeny of family life and social adjustments in various infra-human animals. In *Problems of Infancy and Childhood*. Ed. M. J. E. Senn. New York. Macy.
14. SCHNEIRLA, T. C. 1952. A consideration of some conceptual trends in comparative psychology. *Psychol. Bull.* **49**, 559.
15. SCHNEIRLA, T. C. 1965. Aspects of stimulation and organisation in approach/withdrawal processes underlying vertebrate behavioural

development. In *Advances in the Study of Behavior*. I. Ed. D. S. Lehrman, R. A. Hinde and E. Shaw. New York. Academic Press.

16. THORPE, W. H. 1956. *Learning and Instinct in Animals*. London. Methuen.

17. THORPE, W. H. 1963. Ethology and the coding problem in germ cell and brain. $\mathcal{Z}$. *Tierpsychol*. **20,** 529.

18. TINBERGEN, N. 1963. On aims and methods of ethology. $\mathcal{Z}$. *Tierpsychol*. **20,** 410.

EXPERIMENTAL APPROACHES TO THE EVOLUTION OF BEHAVIOUR*

P. L. BROADHURST

Department of Psychology, University of Birmingham

INTRODUCTION

THE purpose of this chapter is to discuss some of the different approaches to the study of the evolution of behaviour. Specifically, I shall contrast the ethological approach with approaches deriving both from psychology and from psychogenetics. Though a psychologist myself, I tend to favour the last rather than either of the others, and consequently I am somewhat critical of ethology as applied to this problem. But any such criticisms are made in good part and I hope may regarded as contributions to the dialogue going on between these two rather different disciplines, the gap between which is continually being narrowed by a series of excellent bridging works, among them the recent examples by Marler and Hamilton,[20] Hinde[13] and Manning.[19]

Ethologists, perhaps because of their training as zoologists, and because of the concern of part of zoology with systematics, have been especially interested in behaviour as a taxonomic characteristic. That is to say, differences between species and sub-species in various observable aspects of behaviour, especially such items as courting rituals in birds, have proved to be quite as useful for differentiating between species as such morphological characteristics as coat-colour, limb structure, mode of reproduction and so on. The names of Lorenz[15] and Cullen[8] have been especially associated with this view. There is thus a sense in which such interest has resulted especially in

* Based in part on the Hall Memorial Lecture in Ethology given at the University of Bristol in March 1967. Some of the work referred to was supported by U.S. Public Health Service grant No. 08712 from the National Institute of Mental Health. Acknowledgments are due to Professor J. L. Jinks for his comments.

a concentration on evolution *and* behaviour rather than the evolution *of* behaviour, which is my present concern. It was as long ago as 1941 that Lorenz published a classic paper on the use of behavioural characteristics in taxonomic work of this kind in his study of twenty different species of ducks and geese.[14] He showed that some characteristics, such as the way lost chicks give a monosyllabic distress call, are common to all members of the group, whereas some others are common to the ducks only and not to the geese, and some others are characteristic only of one or two particular species. This type of analysis led in some cases to the reassessment of taxonomic positions of certain species, and hence justified the use of behavioural criteria, placing them as equal in value to other, more structural characteristics.

This is not to say that the ethologists have neglected the problem of the evolution *of* behaviour entirely. On the contrary, they have made considerable efforts to understand how particular aspects of behaviour have evolved, and in order to do so have relied largely on the comparative approach. Here again, starting from a zoological background has been of considerable assistance, in that the methods which have succeeded in other comparative endeavours have served well enough here. Thus the ethologists' approach to this problem is necessarily from a strictly biological viewpoint which takes account of evolutionary theory and the nature of adaptive forces at work in problems relating to speciation. While it can be improved in the ways mentioned later, it is probably superior to the efforts which psychologists in general have made to analyze the evolution of behaviour. I will, therefore, consider an example of the type of analysis traditionally employed by psychologists to encompass this problem, before considering a typical ethological approach and finally moving on to the psychogenetic approach which is emerging.

A PSYCHOLOGICAL APPROACH

Of all the workers currently engaged in psychology in this endeavour Bitterman is foremost, because of his recent and sustained attack on the problem.[1] His approach is distinctly superior to the attempts made in the first few decades of the present century by psychologists such as Thorndike,[22] who sought to apply the same experimental test to different species and then to view the results on a comparative basis. In passing, one might mention the apparent neglect by ethologists of these early psychological experiments. Lorenz,[16] in particular, writes as if the Darwinian revolution in the middle of the last century had no effect whatsoever on psychology and that no one in psychology interested himself in comparative problems, evolutionary thought only entering the realm of behaviour studies with the work of Whitman in the United States and Heinroth in Berlin. This is to ignore a large body of work which was concerned with comparisons, fruitless though they may now seem to us, between various species in respect of their so-called intelligence. Even so convinced a behaviourist as John B. Watson[24] made important contributions to the field study of birds: his analysis of the behaviour of Noddy and Sooty terns on one of the small islands, or ' keys ', off Florida in the first decade of the present century is a fine example of careful observation and ingenuity in adapting laboratory methods to the difficult task of field observation. Watson's field work is characterized by patience—not unusual in naturalists, though his report on an eight-hour vigil at one nest may approach a record—and an ingenuity of the sort which is less frequently found and which was decidedly noteworthy considering the epoch. Watson's methods show an inventiveness and freshness of approach which compel attention even to-day.

Leaving aside, then, these early attempts of psychologists concerned with this problem, the results of which led them to follow Thorndike in postulating that the learning process was the same throughout many species, we turn to Bitterman's approach. His shows a degree of sophistication lacking in the early attempts, in that he is concerned to study only aspects of behaviour which can be shown to be within the repertoire of the species selected to represent the various phyla of animals he studies. He has spent considerable time and ingenuity in devising accurate methods for the analysis of a standard series of learning tasks. The basic test situation, which he has adapted for use for the various species, has been a discrimination situation, in which the animal has to make a choice between two stimuli, of which only one is rewarded, usually by food.

The two types of problem on which he has concentrated have been presented both in the spatial and the visual mode. That is to say, the animal may be rewarded on the basis of a spatial discrimination—e.g. always having to choose the left-hand discriminandum, or on the basis of a visual discrimination, i.e. having to learn to choose a particular pattern or colour, no matter on which side of the apparatus it appears. Using these two techniques, Bitterman has used two kinds of learning task. The first is the reversal learning task, in which the positive and negative stimuli are interchanged: as soon as the animal reaches a pre-designated criterion of learning of one of the stimuli as the positive one, it now becomes the negative one, no longer reinforced by food, and the previously negative one becomes the positive. Some animals, e.g. the rat and the monkey, can learn to do this very well, so that they develop a set towards reversals, and in many cases it needs only one trial under the reverse conditions for them to change their discrimination choices completely. Bitterman has been able to show that the species of fish on which he has concentrated (the African mouth breeders) are unable to do this and do not show any such progressive improvement in habit reversal. Thus we have a clear distinction, Bitterman claims, between some of the vertebrate species, such as monkey, rat, turtle, and even the rat when decorticated, and fish and an insect (the cockroach) and an annelid such as the earthworm. The situation is somewhat different, however, between the two modes of presentation of the type of reversal problem, as may be seen from Table I. If a visual instead of a spatial problem is presented, the status of the decorticated rat and the turtle changes, and they are now included in the group which includes the fish, in that they fail to show the progressive improvement in habit reversal characteristic of the rat, the pigeon and the monkey.

The other type of technique which Bitterman has exploited to great effect in the discrimination situation, has been probability learning. If inconsistent reward is applied to one of the stimuli, so that it is positive and rewarded only, say, 70 per cent of the times, the other being rewarded the other 30 per cent of the times, the animal under test can respond in various ways. Typically, the rat maximizes reward, either by responding 100 per cent of the time to the stimulus rewarded for the larger proportion, or displays some probability matching on a non-

TABLE I

BEHAVIOUR OF A VARIETY OF ANIMALS FOR FOUR CLASSES OF PROBLEM WHICH
DIFFERENTIATE RAT AND FISH EXPRESSED IN TERMS OF SIMILARITY TO THE
BEHAVIOUR OF ONE OR THE OTHER OF THE TWO REFERENCE ANIMALS
(Reproduced, with permission, from Bitterman [1])

	SPATIAL PROBLEMS		VISUAL PROBLEMS	
ANIMAL	Reversal	Probability	Reversal	Probability
Monkey	R	R	R	R
Rat	R	R	R	R
Pigeon	R	R	R	F
Turtle	R	R	F	F
Decorticated rat	R	R	F	F
Fish	F	F	F	F
Cockroach	F	F	—	—
Earthworm	F	—	—	—

F = behaviour like that of the fish (random probability matching, and failure of progressive improvement in habit reversal).

R = behaviour like that of the rat (maximizing or non-random probability matching, and progressive improvement in habit reversal).

Transitional regions are connected by the stepped line. The brackets group animals which have not been differentiated by these problems.[1]

random basis, which can be shown to be related to some systematic habit of responding. The fish Bitterman studied, on the other hand, show an essentially random type of behaviour, which issues in probability matching so that it chooses the rewarded stimulus on approximately 70 per cent of the trials, but on a completely random basis which is not related to the pattern of food reinforcement it receives. This random matching, as Bitterman calls it, is shown by the fish, in both spatial and visual discrimination problems, as we can see from Table I. Maximization, as demonstrated by the intact rat, is a very typical result for both types of problem. The pigeon and the turtle, on the other hand, show the rat-type of maximization on spatial probability problems, whereas they display the fish-like random probability matching on visual problems.

Thus Bitterman has been able to begin to construct a sort of phylogenetic scale in which the less complex the organism, the more fish-like the behaviour tends to become. I should add that this pattern is by no means perfect; the status of the work on octopus—an invertebrate—in this respect is, at the moment, in doubt, there being evidence that it does not behave like a fish in that it displays reversal learning,[18] and, moreover, man, as Eysenck[9] has pointed out, is, under certain probability learning conditions, more fish-like than rat-like.[11]   However, Bitterman's is a most interesting attempt to get at the evolution of behaviour; and the choice of psychological techniques has clearly been judiciously made, in that by their use Bitterman has been able to differentiate what he regards as the qualitatively different kinds of behaviour both between and within species.   This technique he calls ' phylogenetic filtration '. However, it should be noted that not all psychologists would accept that Bitterman has succeeded even thus far.   Mackintosh,[17] for example, argues cogently that the phyletic differences are not in fact qualitative but merely quantitative and can be subsumed under general principles of learning.

Moreover, the net outcome of this considerable effort has, as yet, been somewhat disappointing.   Bitterman himself has not dwelt much on the question of the adaptive significance of the change in the type of ' strategic capability ', as he calls it, as we proceed from fish to rats.   Obviously little can be done on the basis of so few species at present represented in this approach—eight only, even if the two sorts of rats Bitterman has used are included.   Thus it seems that it is not yet possible to make generalizations about the evolution of behaviour, and that it will be many, many years before even any class of vertebrates is studied in sufficient detail to provide the data for the inter-species comparisons which could provide a meaningful basis for generalization.   At present, the comparison of species so widely separated in an evolutionary sense is not likely to assist us in any detailed analysis of evolutionary change.

AN ETHOLOGICAL APPROACH

The ethologists' approach in this area, which is essentially one of hybridization, is clearly superior in that it relies on the direct manipulation of the structure of speciation, as may be seen from a recent example from that discipline. The one I have chosen is some work by Sharpe and Johnsgard[21] on the inheritance of behaviour in two species of ducks, the Mallard and Pin-tail. This is a study which carried hybridization to the F_2 generation, i.e. the first-generation hybrids were themselves bred to produce further offspring. This point is of some importance, since it is inevitably unsatisfactory to consider the first-generation hybirds alone. This is because the F_1, on an average, will tend to be uniform, irrespective of the position it occupies on any scale of the particular aspects of behaviour or, for that matter, any other characteristic we wish to measure and in respect of which the parental species are differentiated. Obviously, if the parental species are not so differentiated, it would be unprofitable to study their hybrid offspring, since the most obvious expectation is that in such a case the F_1 would not differ from the parents and so there would be no differences to analyse. Again, in passing, it should be noted that the F_1 hybrid may lie outside the range on the scale defined by the parental values; that is to say, it may display heterosis, or, as it is often called, hybrid vigour. The understanding of heterosis is a complex matter and to pursue it farther would lead us far astray.

Thus it is that the F_2 generation is the first that can reliably be expected to show the sort of segregation of the elements of the behaviour pattern that should prove of interest for analysis. This at once leads to a limitation of the method of species hybridization, since the F_1 hybrids themselves may be infertile and so preclude the possibility of obtaining an F_2. In practice, this limitation means that the method can be applied only to species which differ relatively minimally in terms of their genetic background, since if, for example, chromosome numbers differ, there may be essentially mechanical problems in the production of hybrids at all, let alone the production of fertile hybrids capable of producting an F_2. Sharpe and Johnsgard's study [21] avoided this problem, and they were able to produce a small number of F_2 individuals, which were carefully studied in respect of various aspects of plumage and of various aspects of behaviour. The latter included the responses in the head-up-tail-up complex, including bill-pointing, turning and nod-swimming, as well as the down-up and so-called ' burp ' responses. The results for eleven males of the F_2 generation are plotted in Figure 1, which shows the plumage index scored from 0 to 20 on the ordinate and the behavioural index scored from 0 to 15 on the abscissa. As indicated, the completely Mallard characteristics would give a zero score on both these measures, whereas completely Pin-tail characteristics would yield the maximum score. The correlation between these two indices is 0·76 as indicated by the dotted line.

The conclusions these workers draw from their findings are as follows: first, they suggest these results indicate a " tendency towards the common inheritance of both plumage and behavioural characteristics " and that this is " ample evidence " that they " have a very similar genetic basis ". This is a conclusion which does not follow from the data provided. There is ample evidence, on the contrary, that the two aspects can become widely separated: for example, the individual having a behavioural score of 7 is at least 5 points above what it should be if the two were perfectly correlated. Even allowing for errors of measurement, a divergence of one-third of the whole scale would hardly be expected. The suggestive alignment of the first four points in a direction contrary to what would be expected on the hypothesis of similar genetic basis should also be noticed, though here too errors of measurement may produce such a result.

The second conclusion Sharpe and Johnsgard arrive at is that the results " suggest fairly simple genetic control of each of these characters [the indices used], probably involving relatively few genes ". This seems inherently unlikely and, indeed, is directly contrary to some conclusions reached by Hinde[12] in a study of plumage characteristics in finches and canaries which, he suggests, " depend on multiple factor differences ". But the data themselves also fail to support the authors' conclusion in that there is very little suggestion of two unrelated groups of individuals in this diagram, such as one would expect in the case of the simplest form of genetical control by a few major genes.

Another conclusion reached is that the traits investigated

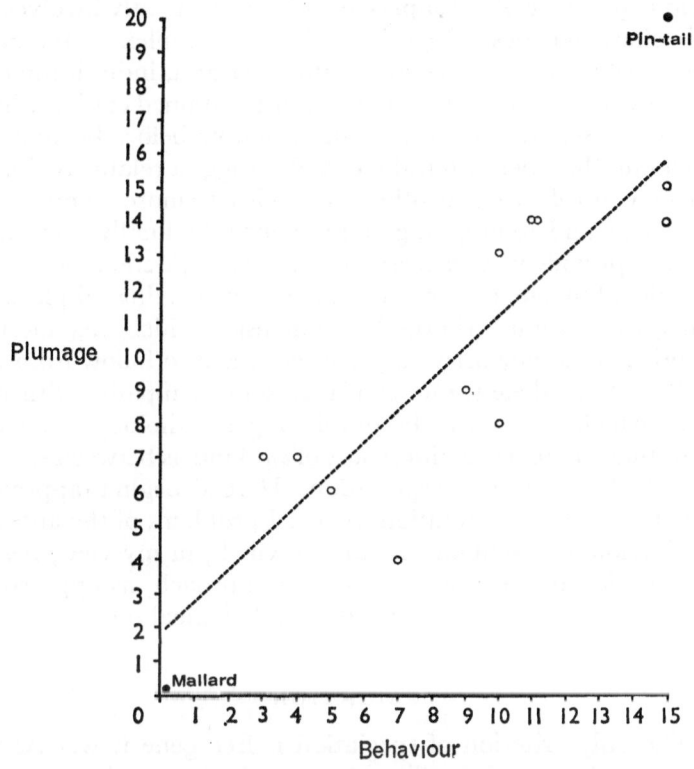

FIGURE 1

CORRELATION OF PLUMAGE WITH BEHAVIOUR

The diagram shows the relation of indices of plumage to behavioural characteristics in eleven individuals of an F_2 cross between Pin-tail and Mallard ducks. Re-drawn from Sharpe and Johnsgard.[21]

are probably of significance as isolating mechanisms between species and should therefore be sensitive to rapid modification under the pressure of natural selection. But, here again, this does not seem to follow from the previous findings, since, if such were the case, the characteristics in question should be so

C

tightly linked genetically that they should segregate as blocks, rather in the way that species which are polymorphic do. In such cases, two extreme forms are to be observed, and the intermediate forms do not occur and are at a selective disadvantage. So called ' super-genes ' are frequently involved in such characteristics. In this work, the complex of plumage and of behaviour does *not* act in this way and, indeed, some of the individual aspects of the behaviour mentioned earlier, which were rated separately by these investigators before being combined into their behavioural index, did suggest relatively simple genetic control, whereas others were clearly more complex.

Sharpe and Johnsgard go on to conclude, finally, that their results " provide further reason for seriously questioning the use of male plumage patterns or male pair-forming displays as primary taxonomic criteria ", a conclusion which, like most of the other ones they arrive at, does not seem to follow from the results. Nevertheless, one can have some sympathy with this view—which appears to be developing in ethology—that the magnitude of observed differences of *any* kind as between species is not the best criterion of speciation. Here we begin to approach what I regard as the solution to vexed problems of the analysis of behavioural evolution, a solution which, in my view, lies in the adoption of a specifically genetic approach, as opposed to a zoological or, still less, a strictly psychological one.

SPECIATION

The only criterion of speciation is free gene flow. As we have already seen, infertility between similar species bespeaks a recent divergence in respect of evolution, and it is only the genetic content that is the ultimate criterion upon which all others must be based. Differences of any kind, however interesting, are not really relevant to the argument. A simple example may serve to convince. Within any single species a mutation, for example albinism in the Norway rat, can give rise to gross differences between the mutant and the non-mutant form. Yet these two forms are completely inter-fertile and there are no problems whatsoever in respect of mating and production of viable offspring. Yet the differences observed, both in morphology and in some aspects of behaviour, may be just as great as those between species, and used for making a crucial taxonomic decision. If this view is accepted, then the standard techniques of genetics can be employed to throw light on the evolution of behaviour, just as they can, indeed, on any other phenotype measured. For this purpose, we do not need inter-species hybrids, since the elucidation of the genetic architecture of a single species and the gene action involved can give valuable clues as to the evolutionary forces which have operated on that species alone. Thus, genetics is evolution and vice versa; only behavioural analysis firmly based on genetics can hope to supply the answers to problems within and between species. In this way, we can at once side-step all the difficulties involved in hybridization and cut across all the taxonomic arguments to which they inevitably give rise.

Moreover, the techniques of genetics in general, and bio-metrical genetics in particular, deal not in terms of individuals, but in terms of populations. We are, therefore, more con-cerned with population means and variances, and in this way the statistical techniques of reliability and validity so dear to psychologists can be brought to bear on the problems of measure-ment. Hybridization, typically, does not lend itself to precise measurements, since it must be largely conducted in wild or semi-wild conditions. Within-species analysis, on the other hand, can be conveniently conducted using large numbers and within the laboratory. A singular advantage of laboratory work in this connection is the meticulous control over the en-vironment of the organisms to be studied which can be achieved from the period even prior to its conception right through to the time of measurement by whatever test is required to establish the phenotype to be measured. This control cannot, in the nature of things, be obtained under naturalistic conditions and, in consequence, such observation is limited to those aspects of behaviour for which it is possible to assert unequivocally that there is little environmental determination. Clearly, if the behaviour to be studied is modifiable to any great extent by, for example, learnt responses, or dietary influences, or tempera-ture and so on, then lack of knowledge of the extent of these influences may vitiate the results to an important degree.

Hence ethologists have concentrated on responses which they regard as innate and largely unmodifiable by experience. In the laboratory, on the other hand, most phenotypes can be assessed, not only because the environment can be controlled therein, but also because knowledge of the environment makes possible the inclusion of environmental factors in the analysis. In this way, traits having relatively low heritability are not ruled out for that reason alone, and can be analysed with equal facility.

A PSYCHOGENETIC APPROACH

For the psychogenetic approach, therefore, we use, not species, but strains or varieties within a particular species. The understanding of what goes on within a species in a precise genetical sense probably precedes in importance the inter-species differences and speciation as such. Obviously, a difference between the parental strains in respect of the phenotype it

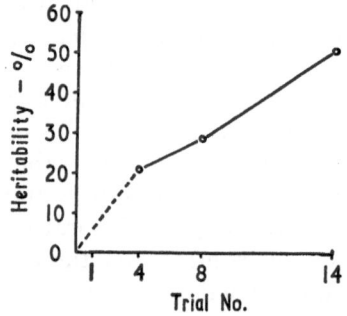

FIGURE 2

RUNNING TIME IN MICE

The figure shows the increasing herit-ability of maze running time as a function of practice. Derived from data provided by Vicari.[23]

is desired to measure must obtain, but once such a difference— or differences between more than one pair of such strains—is obtained, than the analyses can proceed with relative ease. This type of work, which is called ' behaviour genetics ' in the United States,[10] but for which the term ' psychogenetics ' is to be preferred, has now arrived at a stage which allows some speculative comments about the evolution of behaviour.

As yet, however, there are few examples of the fruits of this type of approach. Nevertheless, let us consider what there are. Figure 2 presents a re-analysis of some old data collected many years ago by Vicari [23] which relate to the speed of maze-running of mice. She reported a large-scale study using four inbred strains and gave the scores for the various generations for successive phases of the learning experiment. This enabled us to calculate [6] the change in the heritability index of one of these measures, the mean running time in the apparatus. On the first trial the value was not significantly different from zero, but it rose regularly as the experiment progressed. This suggests that the increasing heritability here represents the progressive release of a genetically determined response from the effects of environmental stimuli which are irrelevant to it but are apt to obscure its action in the early stages of the training. Thus, even within the compass of a few weeks of training in a learning experiment, there is an interaction between environment and heredity of the sort which, on the large scale, in terms of numbers of animals and whole aeons of time, gives rise to the evolutionary changes recognized as speciation. Another example is to be found in the work of Fuller and Thompson [10] on behaviour in dogs. They found that, in a variety of tests, the heritability fluctuated apparently randomly as the number of trials increased.

Some recent work [7] has carried this approach a stage farther. Extensive use has been made of an analysis of the behaviour of rats in the open-field test of emotionality. In this test rats are exposed for two minutes on four successive days to a mildly stressful situation, to which they respond by ambulating around the arena in which they are confined and indulging in emotional elimination, especially defecation.[2] Six different strains of domestic rat were used which were crossed in all possible combinations to yield a 6×6 diallel table of results.[3, 4] This methodology [5] has been developed by the Birmingham geneticists for use with plants, and has proved extraordinarily well-adapted to the problems encountered with behavioural phenotypes which, curiously enough, quite closely resemble those encountered in plant breeding. This type of analysis, like the Vicari results in mice, allowed us to evaluate [7] the importance of the environmental effects, and Figure 3 shows the components of variation for the ambulation response in the open-field test on the four successive days of testing. The

bar diagrams indicate the size of the respective components—
E_2 is the measure of the environmental differences between
litters, D_R is the component ascribable to additive genetic
variation and H_R is that relating to the dominance variation.
As before, across the top the heritabilities are shown as a percen-

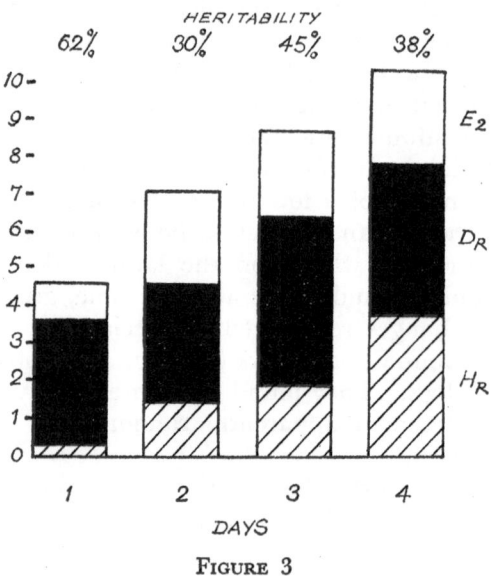

FIGURE 3

COMPONENTS OF VARIATION FOR AMBULATION

The bar diagrams indicate the size of the components of variation derived from the ambulation scores in the open-field test of six strains
of rats on four successive days. Heritability of
the measure, calculated, in the 'narrow sense',
as $\frac{1}{2}D_R/(\frac{1}{2}D_R + \frac{1}{4}H_2 + E_2)$, is shown above.
(Based with the permission of the Royal Society
on a figure in Broadhurst and Jinks.[7])

tage. The additive component, in solid black, is virtually constant throughout, whereas the dominance component increases
somewhat, first achieving a significant difference from zero on
Day 2 and retaining it thereafter. The estimate of an environmental variation (E_2) increases somewhat, with the result

that the heritability as shown declines with experience, instead of increasing (as we saw in the maze-learning of the mice).

We can carry this type of analysis a stage farther by evaluating the proportion of dominant to recessive alleles carried by each of the six parental strains, as shown in Figure 4. On the ordinate we have a symbolization which implies the higher the numerical value the greater the proportion of recessives. The abscissa gives the actual ambulation score for the strain as obtained from the mean of the appropriate array in the diallel table mentioned earlier. There are four entries for each of the six strains and the arrows indicate the progressions from Day 1 to Day 4. Thus we can see that on Day 1, for example, strain 2 contains most dominants, and, moreover, has the lowest score, while strain 6 contains the fewest dominant genes, and has the highest score. This situation, as you can see, changes gradually over the four days, until, if we consider Day 4, it is the two strains which now have the most extreme values for dominance (5) and recessives (4) which have the intermediate ambulation scores, while the parents with the extreme scores, such as strain 2, have become more intermediate with respect to the proportion of dominant to recessive alleles determining the behaviour. This further example of the change in the genetic structure controlling behaviour during the fairly minimal experience involved in successive exposures in the open-field test, allows us to make certain inferences about the natural selection which must have taken place in giving rise to this type of behaviour in the rat. Given the demonstrated increase in dominance as shown in Figure 3, and the emergence of an intermediate scoring strain as one with the greatest proportion of dominant genes, it becomes possible to speculate that it is the intermediate behaviour, in terms of ambulation score, that is presumably favoured by natural selection. This makes reasonably good adaptive sense. If we consider the natural habitat of the wild rat, a nocturnal animal subject to predators both on the ground and in the air above, it is clear that too much exploration in the open in a strange situation might well imperil the species. On the other hand, too little might be equally dangerous in the sense that it could result in available food supplies remaining undiscovered. It is in these terms that our finding—that an

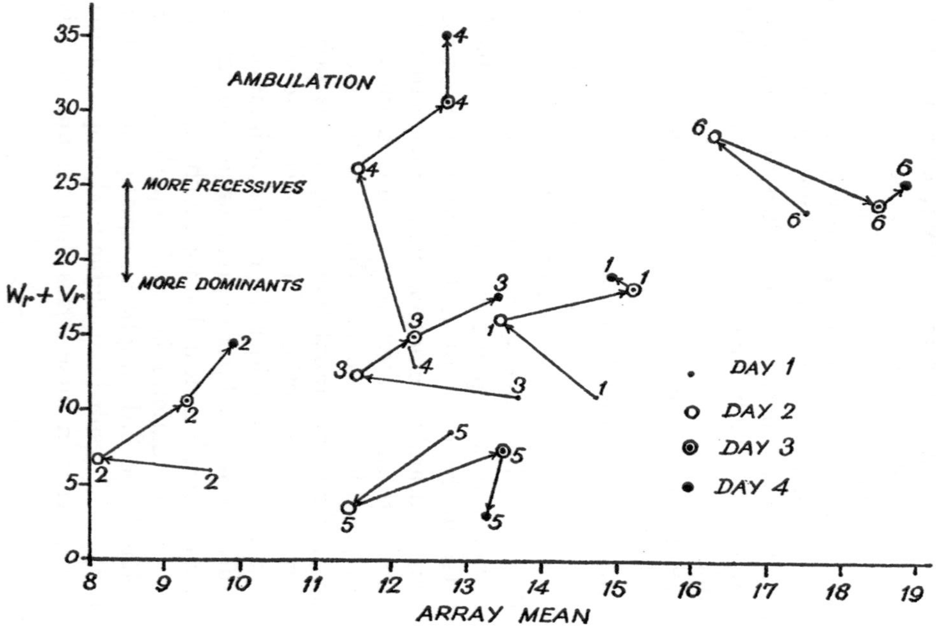

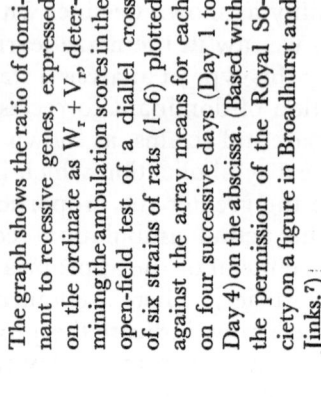

FIGURE 4

RATIO OF DOMINANTS: RECESSIVES
AS A FUNCTION OF AMBULATION

The graph shows the ratio of dominant to recessive genes, expressed
on the ordinate as $W_r + V_r$, determining the ambulation scores in the
open-field test of a diallel cross
of six strains of rats (1–6) plotted
against the array means for each
on four successive days (Day 1 to
Day 4) on the abscissa. (Based with
the permission of the Royal Society on a figure in Broadhurst and
Jinks.[7])

intermediate amount of exploration appears to confer the greatest fitness—can best be understood. The problem as to why this relationship should not be apparent from Day 1, but emerge only at successive daily trials, has recently been solved by Whimbey and Denenberg[25] who showed, by factorial analyses of behaviour—including open-field ambulation—that the func-

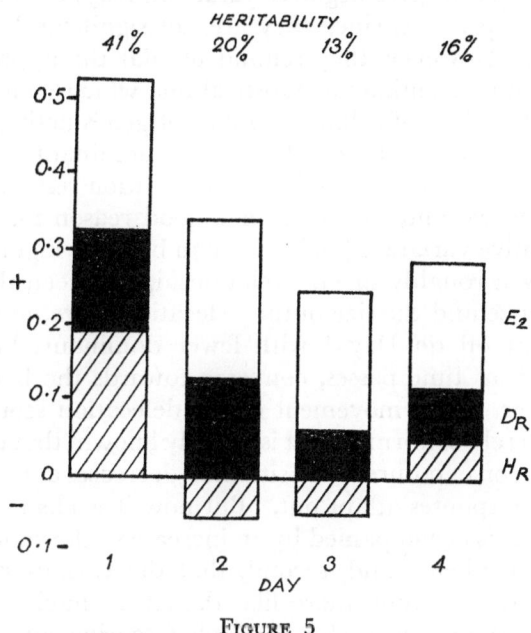

FIGURE 5

COMPONENTS OF VARIATION FOR DEFECATION

See Figure 3 for details. (Based with the permission of the Royal Society on a figure in Broadhurst and Jinks.[7])

tion it subserves on Day 1 differs markedly from that on Day 2 to Day 4.

If we now turn to the other type of behaviour observed in the open-field test of emotionality in the rat—emotional elimination, or defecation—a totally different picture emerges. This other behavioural phenotype was also subjected to biometrical analysis in respect of day-to-day variation. Figure 5 shows the bar diagram for the components of variation for defecation, in precisely the same terms as before. Here we can see a steady reduction in the magnitude of the additive component of variation (D_R) which, indeed, is significantly different from zero only on Day 1. This change is reflected in the decrease in the values indicated at the top for heritability. The dominance component (H_R) shows small erratic fluctuations, and indeed gives negative values on Days 2 and 3, which need not surprise us, since they are not significantly different from zero. However, they remind us that the approach employed here is essentially a statistical one which endeavours to estimate the effect of a large number of genes acting cumulatively. If we now consider the ratio of dominant to recessive alleles of those genes controlling the defecation response, shown in Figure 6, we find that there is a good reason for a decline in the additive variation (D_R), shown in black in Figure 5. The plots show a roughly inverse relationship between the degree of dominance and the size of the defecation score, in that most strains start off on Day 1 with fewer dominants but higher scores, and, as time passes, converge towards the bottom left-hand corner. This movement of the defecation scores on the abscissa merely confirms, what is already known, that emotional elimination on exposure to the situation decreases as the animal's emotional responses adapt out. But now it is clear, first, that this decrease is accompanied by an increase in the contributions of dominant alleles, and, second, that the various strains are progressively becoming more like the strain having the most dominants—which was also the lowest scoring one, first and last. Clearly, dominance in this situation is for low defecation in the open field, and the successive days of testing provide an environmental stimulus which increases the expression of those genes responsible for low defecation. Moreover, the movement among the higher-scoring strains, in respect of the ratio of dominant to recessive alleles, suggests that those high-scoring strains which respond rapidly to the test conditions are at a selective advantage over those which respond either less or more slowly, such as strain 3.

It is important to be vigilant in seeking artefacts in such analyses as, indeed, we must be in any behavioural measurement. The possibility remains that our interpretation of this

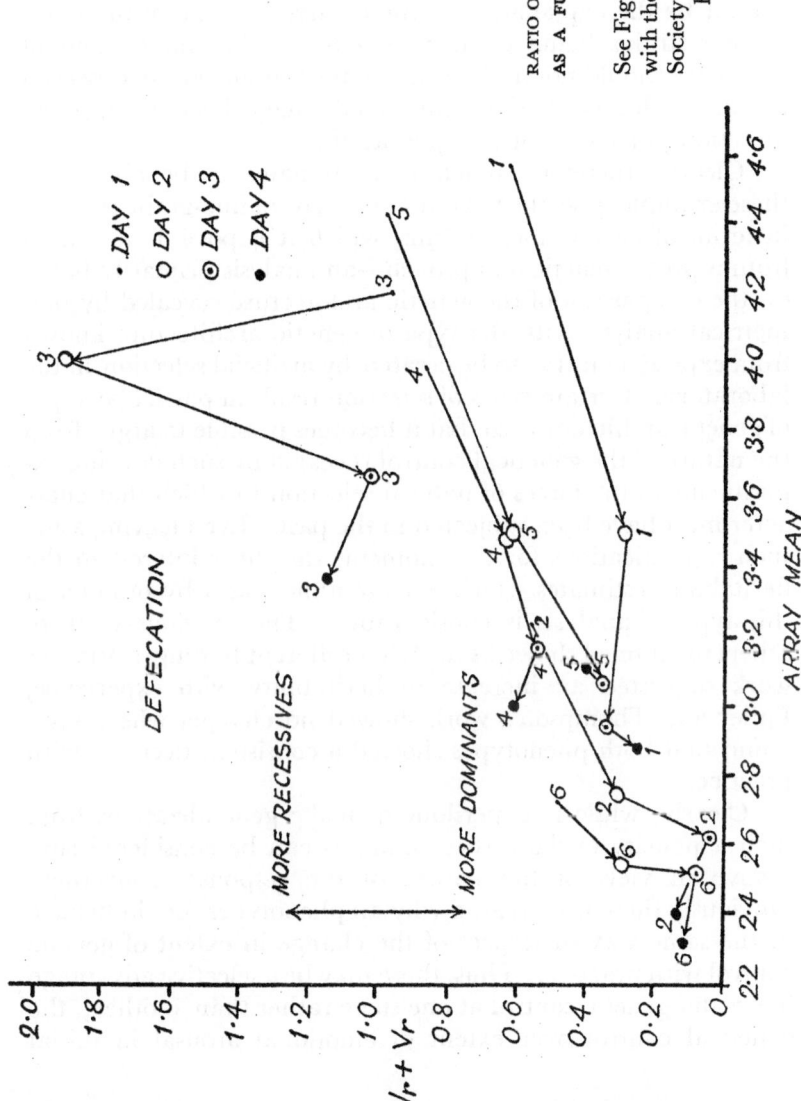

change over days might be vitiated by the fact that the behavioural change detected, namely the reduction in low-scoring strains, must necessarily be less than for high-scoring ones as soon as zero is approached—negative behavioural scores cannot of course, be admitted, even though we can tolerate statistical ones occasionally, as in Figure 5! However, this cannot be the explanation of the change in some of the high-scoring strains being greater than it was in others; and of course we should not underestimate the significance of observed trends of this kind, since they could have been in opposite directions, or have not emerged at all.

Clearly, therefore, much more remains to be done, but these examples give the flavour of the type of fine-grain analysis, in terms of genetic mechanisms, which it is possible to obtain from a psychogenetical approach—an analysis primarily based on the comparison of the genetic architecture revealed by bio-metrical analysis with the type of genetic architecture known from experimentation to be created by artificial selection in the laboratory. Certain types of selection result in particular types of genetic architecture, so that it becomes possible to argue from the nature of the genetical control revealed by such crossing experiments to the forces of natural selection to which that character must have been subjected in the past. Even leaving aside such sophistications for the moment, the sheer interest in the heritability estimates which emerge almost as a by-product of this type of analysis is considerable. The previous work of others mentioned showed somewhat different findings: Vicari's work suggested an increase in heritability with experience, Fuller and Thompson's work showed no change, whereas we found that both phenotypes showed a consistent decrease with practice.

Clearly, while it is perilous to make generalizations from such a paucity of data, these findings can be considered suggestive in view of the nature of the responses concerned. Obviously there is no reason why all phenotypes should behave in the same way in respect of the change in extent of genetic control with practice. Thus, there may be a selective advantage in greater genetic control at one stage rather than another; the genetical control over extent of emotional arousal in initial exposure to strange situations like the open-field test may be more important than in subsequent exposures, during which the extinction of fear is progressing; in a learning situation, like Vicari's, on the other hand, the selective advantage may lie in the increased genetic determination in the later stages, in which the animal is typically able to solve problems, such as those posed in maze learning, with greater speed and fluency. These are only speculations, but they do indicate the power of the methodology in generating such hypotheses, which clearly are now open to rather precise testing in laboratory situations of various kinds.

This type of analysis takes us some way further than either the rather crude comparative approach of the psychologist, or the more sophisticated, but still unsatisfactory, species hybridization approach of the ethologist. The psychogenetic approach combines the techniques both of the psychological analysis of behaviour and of the genetic analysis of gene action in the most intimate way possible. It is, of course, universally applicable to all phenotypes of living organisms, specifically not excluding man himself. The limitations imposed by the impossibility of breeding man as an experimental animal hamper us considerably. But the natural experiment of twinning, and the analyses of parental and other familial correlations bid fair to allow some attack to be made on phenotypes of direct interest to man himself, and to the understanding of his own evolution. Thus the approach is completely general and seems to hold much promise for the future.

REFERENCES

1. BITTERMAN, M. E.   1965.   Phyletic differences in learning.   *Am. Psychol.* **20,** 396.
2. BROADHURST, P. L.   1957.   Determinants of emotionality in the rat : I. Situational factors. *Br. J. Psychol.* **48,** 1.
3. BROADHURST, P. L.   1959.   Application of biometrical genetics to behavior in rats. *Nature, Lond.* **184,** 1517.
4. BROADHURST, P. L.   1960.   Experiments in psychogenetics : Application of biometrical genetics to the inheritance of behaviour.   In *Experiments in Personality* : Vol. I, *Psychogenetics and Psychopharmacology* (p. 1) Ed. H. J. Eysenck. London. Routledge and Kegal Paul.

36 INFLUENCES ON BEHAVIOUR

5. BROADHURST, P. L. 1967. An introduction to the diallel cross. In *Behavior-genetic Analysis* (p. 287) Ed. J. Hirsch. New York. McGraw-Hill.
6. BROADHURST, P. L. and JINKS, J. L. 1961. Biometrical genetics and behavior : Re-analysis of published data. *Psychol. Bull.* **58,** 337.
7. BROADHURST, P. L. and JINKS, J. L. 1966. Stability and change in the inheritance of behaviour in rats : A further analysis of statistics from a diallel cross. *Proc. R. Soc., B.* **165,** 450.
8. CULLEN, J. M. 1959. Behaviour as a help in taxonomy. *Publs. Syst. Ass.* No. 3, 131.
9. EYSENCK, H. J. 1967. *The Biological Basis of Personality.* Springfield, Ill. C. C. Thomas.
10. FULLER, J. L. and THOMPSON, W. R. 1960. *Behavior Genetics.* London. Wiley.
11. GRANT, D. A., HAKE, H. W. and HORNSETH, J. P. 1951. Acquisition and extinction of a verbal conditioned response with differing percentages of reinforcement. *J. exp. Psychol.* **42,** 1.
12. HINDE, R. A. 1956. Breeding success in cardueline interspecies pairs, and examination of the hybrids' plumage. *J. Genet.* **54,** 304.
13. HINDE, R. A. 1966. *Animal Behaviour: A synthesis of Ethology and Comparative Psychology.* London. McGraw-Hill.
14. LORENZ, K. 1941. Vergleichende Bewegungsstudien an Anatinen. *J. Orn. Lpz.* **89** (Suppl.), 194.
15. LORENZ, K. 1950. The comparative method in studying innate behaviour patterns. *Symp. Soc. exp. Biol.* **4,** 221.
16. LORENZ, K. 1958. The evolution of behaviour. *Scient. Am.* **199,** 67.
17. MACKINTOSH, N. J. 1967. Comparative psychology of reversal and probability learning. In *Discrimination Learning.* Eds. N. S. Sutherland and B. Gilbert. London. Academic Press.
18. MACKINTOSH, N. J. and MACKINTOSH, J. 1964. Performance of *Octopus* over a series of reversals of a simultaneous discrimination. *Anim. Behav.* **12,** 321.
19. MANNING, A. W. G. 1967. *An Introduction to Animal Behaviour.* London. Arnold.
20. MARLER, P. R. and HAMILTON, W. J. III. 1966. *Mechanisms of Animal Behavior.* London. Wiley.
21. SHARPE, R. S. and JOHNSGARD, P. A. 1966. Inheritance of behavioral characters in F_2 mallard × pintail (*Anas platyrhynchos* L. × *Anas acuta* L.) hybrids. *Behaviour* **27,** 259.
22. THORNDIKE, E. L. 1911. *Animal Intelligence.* New York. Macmillan.
23. VICARI, E. M. 1929. Mode of inheritance of reaction time and degrees of learning in mice. *J. exp. Zool.* **54,** 31.
24. WATSON, J. B. 1908. The behavior of Noddy and Sooty terns. *Publs. Carnegie Instn.* **103,** 187.
25. WHIMBEY, A. E. and DENENBERG, V. H. 1967. Two independent behavioral dimensions in open-field performance. *J. comp. physiol. Psychol.* **63,** 500.

THE ETHOLOGICAL STUDY OF MAN *

D. G. FREEDMAN

The Committee on Human Development, University of Chicago

THE ethological study of man means, simply, looking at man's behaviour through evolutionary glasses. On the face of it, there should be no reason why studies of man could not enter into the mainstream of ethology. However, taking the International Ethology meetings as an indication, we find that over the past six years only three papers have been given whose subjects were hominids. This lack has been regretted by ethologists and at the 1967 meetings a committee was formed to encourage such work. I am not quite clear why there has been this reluctance to work with man.

Perhaps one obstacle in the way of the consideration of man's behaviour in evolutionary terms is that his behaviour is said to be almost entirely culturally determined. Furthermore, it is claimed by some (e.g. White[51]) that biology may be ignored, since culture evolves via its own set of rules. In a similar vein, such well-known biologists as Huxley[29] and Waddington[49] have suggested (following Darlington[9]) that cultural transmission is a new form of information transfer; in other words, culture is a new step in the evolution of genetical systems. Although there is some truth in this view, they have tended to distract us from the job of extending Darwin's insights[10] by a careful consideration of the adaptive value of various aspects of human behaviour. Thus, I prefer to emphasize that no cultural development in man is independent of his biology, and that at some level of analysis *everything man does has an evolved biological component.*

This cultural versus biological dichotomy is further obscured by the perennial opposition of the innate and the acquired. The problem becomes clearer when we consider that the common ethological meaning for innate is ' unlearned '.

* This material has also appeared in ETKIN, W. (Ed.), *Social Behavior from Fish to Man.* Univ. of Chicago Press, 1967.

Is imprinting, the quick formation of attachments between adult and young precocial birds, innate (unlearned) or acquired (learned)? If this question is pursued at any length, we run into logical difficulties. Imprinting, it will follow, involves the unlearned capacity to learn within a short span of time; that is, temporal quickness is given as structure whereas learning is presumably without structure. There is, however, no logical dividing line separating structure from learning, since learning itself must be a structural product of evolution, and the dichotomy comes to a halt at an insoluble impasse. Nevertheless, while we may thus become logically enmeshed with the term 'innate', in other instances, e.g. the patellar reflex, its use is perfectly clear.

What are we to do about these vocabulary problems? It makes considerable sense, when dealing with complex behaviour like imprinting, to speak, at a more abstract level, of an *evolved* capacity to imprint. ' Evolved ' refers only to the fact that imprinting has a clearly perceived adaptive function, but makes no implication about whether or not learning plays a role; it does not suggest opposition with ' acquired ', as does ' innate ', but simply denotes probable phylogenetic origin.

Thus, when we refer to evolved or phylogenetically adaptive behaviour we are designating a behavioural unit which has been produced by the evolutionary process in much the same way that physical or biochemical characteristics of the species have been produced. We are *not* offering an analysis of the complex interaction of gene and environment that went into the development of the behaviour in question, any more than we make such an analysis when we speak of a physical characteristic such as the pink colour of the flamingo. This colour is properly discussed as an adaptive characteristic of the species, even though experiment has shown it is expressed only under particular nutritive conditions. In the same way, a particular behaviour is properly examined from the evolutionary or ecological viewpoint as an evolved unit, irrespective of whether the particular genetic and environmental interaction necessary for its expression has been analysed or not.

By many ethologists, who have become accustomed to opposing innate to acquired, insistence on substituting the inclusive terms ' evolved ' or ' phylogenetically adaptive ' is not seen as an improvement. Lorenz, in dealing with recent objections to the use of ' innate ' by ethologists themselves, has attempted to salvage the term by posing intercalated chains consisting of learned and innate segments.[33] It becomes extremely difficult, however, to imagine such a chain in dealing with, say, the human smile. The fact is that isolation experiments are usually necessary to establish innateness,[33] but in the highly social primates we know that isolation is entirely antithetical to normal development,[1] and the experiment simply cannot be done. At best, we can pose a spectrum of behaviours ranging from reflexes—which are clearly unlearned—through the fixed action-patterns of lower animals—in which Lorenz's intercalated chains of innate and acquired segments may be possible—to such phylogenetically adaptive hominid mechanisms as playfulness, smiling and laughter, which are thoroughly and inextricably enmeshed with learning and experience.

It may be anticipated that a difficulty concerning the term ' evolved ' is that it is frequently used to describe cultural as well as phylogenetic evolution, and the term ' phylogenetically adaptive ' may therefore be the better of the two. For the purposes of the present paper, however, these two terms will be used interchangeably.

ADVANTAGES AND PROBLEMS IN STUDYING ONE'S OWN SPECIES

It is clear that each animal species has its own system of signals and is especially attuned to attend to them. Man is no exception; we are, as conspecifics, constructed to comprehend each other. Given the proper setting and rapport, we can, for example, quite completely immerse ourselves in the nuances of another person's feelings and thoughts. We cannot do this with other animals to the same degree (although we and dogs have become highly attuned to each other via the genetic selection involved in their domestication).

In studying man, if we remain strict behaviourists and deal only with our sense perceptions—as when dealing with inanimate objects or lower animals—we deprive ourselves of

D

insight into vast areas of human experience to which we are naturally attuned, such as affect imagery and thought. Thus, in a discussion of the significance of dance and music for a human group, it is a poor ethnography indeed which neglects an explicit appreciation of the joy felt by the participants and the feelings of communion within the group.

In this regard, the animal behaviourist, of course, is not so well off as the student of human behaviour. While it is true that ethologists have been interested in the experienced world, or *Umwelt*, of their subjects, following von Uexküll[48], their only recourse has been to attempt its reconstruction via objective experimentation (Tinbergen [47a]).

One can, on this basis, criticize Darwin's pioneering work, *The Expression of the Emotions in Man and Animals*; for it is clear that he refrained from giving more than a perfunctory description of the subjective aspects of the emotions, preferring instead to describe in detail only the observables. It is interesting to note that Darwin also refrained from speculating on the adaptive value of man's emotions, perhaps also in the interests of maintaining ' objectivity '. Given the fact that those were pre-psychoanalytic, pre-phenomenological days, and that the very consideration of man's emotions as species-specific behaviour was a major step forward, we can only say of Darwin, as Lorenz has said, that he " knew more than he was able to say ".

While this paper makes some attempt to rectify the scientist's tendency to discuss only the *without* of things, it will fall far short of the requirements of a phenomenological approach as stated, for example, by Buytendijk.[8] It is hoped that, in the future, works will appear which combine in a more thorough manner an evolutionary approach with phenomenological analysis.

I should like now to engage in a brief discussion of various aspects of human behaviour, as seen through the eyes of this evolutionist. The first topic will be sexual dimorphism.

SEXUAL DIMORPHISM

Much has been written contrasting men and women, boys and girls, but such differences have rarely been considered within an evolutionary context. The many discussions of the

Oedipus complex, and how boys and girls differentially master it, provide a good example; we know of no such discussion which seriously considers the obvious parallels between human male-male rivalry, which starts among juveniles, and similar behaviour seen in other social species (e.g. the rhesus, Harlow and Harlow [27]).

From the work of Young,[54] Harris[28] and others we know that the introduction of androgens to young female rats and to embryonic female monkeys permanently virilizes them and, for example, increases their aggressiveness and the number of challenges to fight. It now appears that it is the action of these androgens on the central nervous system during a critical period which produces maleness, and studies of male pseudo-hermaphrodites have yielded similar conclusions regarding the formation of human sexuality (Landau, 1966, personal communication).

It now seems likely that the upsurge of feelings of rivalry which male human four- and five-year-olds and juvenile primates experience are due in some part to shifts in hormonal levels (most probably androgens), acting upon a virilized central nervous system. The compelling urge to win and be ' top dog ' seems predicated upon the evolution of social dominance, so widely seen among group-living species; and in this regard the infantilized human seems to differ from other primates mainly in the time scale of development. More generally, we cannot persist in the notion that behavioural consequences of hormonal differentiation of the human sexes occur for the first time at puberty, as some writers, (e.g. Ausubel,[4]) have held.

Although their explanations have ignored evolution, psychologists have gathered a good deal of data on male-female differences. These findings seem always to reflect greater passivity in females and greater aggressivity in males. Even in cultures where women are more active in courting (the Navajo, for example), it is acknowledged within the culture that the males are saving themselves for heroic bursts of energy. There are studies[4] which show that boys anger more easily than girls when frustrated either by the activities of others or by the resistance of the inanimate materials they are working on. Young males engage in more rough-and-tumble play, are more highly investigative and intrusive, and seem better able to manipulate mechanically and to visualize three-dimensional relationships.[6] Save for the last point, young primate males and females are differentiated in much the same way.

At about seven years of age human males are more given to forming competitive hierarchies than females; they are more interested in assuming the hero's role, as is demonstrated by their competitive behaviour in sports and in their active imagination and day-dreaming. Girls seem more oriented towards their adult role as mothers and they seem everywhere to indulge in play which emulates the caretaking role of women. As adolescence approaches, girls participate more and more vicariously in the world of male competition; they seem not to seek the hero's role but are instead adulators of the hero. Studies of adolescent dreams in various cultures bear this out and show also that the dreams of adolescent boys have more aggressive content than those of adolescent girls.[7] Much of the information on differences between males and females is summarized in Maccoby.[35]

At puberty the great increase of androgen level in males makes that period particularly touchy in terms of male-male aggression and rivalry. This fact may well be the biological underpinning for puberty rites and male initiation ceremonies, for the rivalrous newly adult males must somehow learn the rules which have been devised to prevent aggression from fragmenting the group.

With regard to pubescent changes in women, menstruation now starts earlier than ever before in history,[47] most probably as a result of improved nutrition and the elimination of many diseases. The flexibility of menstrual age seems to have evolved with the apparent function of limiting or extending the fertile years of females according to the availability of nutriments, which in turn serves to diminish competition for limited food sources. Many analogous examples in other species, in which reproductive potential is adjusted to ecological conditions, are described by Wynne-Edwards.[53] Human females also mature more quickly than males in all phases of development, such as bone age, teeth eruption, language development, and sexual maturity; at least one adaptive function of this disparity is that it makes it easier for the males to maintain the just-matured females in a submissive posture since females are younger and less experienced than males at a comparable level of sexual maturity.

While a modern female may bristle at the suggestion that evolution has arranged for her submissiveness, it appears that dynamic dominance-submission relations are the only reasonable means of achieving social stability, given the concurrent adaptational value of a high level of male aggressiveness. On this point, the relatively limited degree of dimorphism among hominids makes it possible for the female to challenge male dominance frequently, a fact of hominid life we usually call the ' battle of the sexes '. Such challenges are almost always verbal or indirect, and many societies reduce such intersexual conflict by institutionalizing completely separated roles. The psycho-analytic concept of ' penis envy ' is based on similar data, and translates very readily into such a biocultural framework.

Male beardedness and overall hairiness seem best explained in terms of both male-male competition and male-female non-competition. Young men, though strong, are relatively hairless and to some extent the hierarchy based on age is partially stabilized by the tendency to defer to signs of age, including beardedness and hairiness. Awe and respect for the older seem built into man,[42, 49] and beardedness as well as greying may well have evolved as means of designating position in such an arrangement. Among some beardless peoples, such as the Maori, it is interesting that facial tattooing or scarring became a common practice, with the tattoos and scars much more extensive and " frightening " on the males, more decorative on the females. This may, in fact, give us some insight into the function of beardedness, for (as current research by the author suggests), the bearded face, too, takes on a forbidding or authoritative appearance more readily than does a non-bearded one. This is in keeping with what is frequently found in other species —that strong dimorphic traits correlate with the establishment of male dominance hierarchies.

Lest human male-female differences receive over-emphasis, let us again note the moderate physical dimorphism exhibited in Hominidae compared with other primates and the fact that a good deal of exchange of roles, except in terms of warfare and heavy work, is seen in most culture areas. The attempt to obliterate the evolved differences between the human sexes[38] is however, by far the more prevalent distortion to be found in recent thought.

COURTSHIP

Parental care of the young and maintenance of groups through family ties seem to be the major organizational functions served by permanent pairing. In mankind the probable function of extended pairing is to assure successful rearing of young within a minimal group. Thus wolves need the pack for the hunt, ungulates need the herd for protection, and most primates are found in sizeable groups, probably as protection against predators. Man, endowed with his intelligence, is probably viable at the level of family groups, and prolonged attachments between mating pairs is characteristic. In this section, then, we will consider the means by which males and females first unite.

What are some of the more apparent attributes which attract men and women to one another? In general, men are more readily sexually excited by the visual modality than are women, whereas women first react with comparable intensity when physically contacted.[31, 37] A similar dimorphism is found in a majority of species, the male usually taking the active courting role on catching sight or smell of the female. This may be followed by ritualized movements and then by the actual touching of the female's genitalia in a mount. In animals ranging from Drosophila[45] to the apes,[12] the female's major sexual activity is often restricted to rejection of unwanted suitors.

In each such species, then, there are visual, and often odoriferous, attributes of the female to which the males are attracted. What, then, are the attractants exhibited by the human female? The entire concept of female beauty is here our subject matter, and so a complete discussion is not realistic. It is interesting, however, that such a frequently discussed subject has rarely been considered in an evolutionary context.

Regarding the so-called culture-boundness of beauty and

disregarding for the moment the obvious fact of gene-pool differences with their associated physical differences, we observe that men the world over seem to have approximately the same taste in female beauty. The marauding armies of history are perhaps the best proof of this. While one often hears that the male Hottentots, for example, prefer women who have grotesquely large buttocks (following Darwin[10]) the fact is that young Hottentot girls have buttocks which are simply well-rounded and attractive by almost anybody's standards. Only as the women grow older and fat accumulates do they look grotesque to other peoples; but almost all women lose their sexual appeal as they age beyond their reproductive years.

Softer facial features and relative hairlessness of the face and body characterize human females, and Lorenz has spoken of childlike non-angular facial features associated with facial fat, which are retained in postpubescent females but not in males. Similarly, female musculature is softer, giving the entire body a more supple, relatively non-resistant appearance; along with this the relative smallness of females emphasizes their non-competitive submission.

The fact that the human female is the only mammal whose breasts are conspicuous whether or not lactating and the fact that breast size and milk production are not related, coupled with the clear attractiveness of breasts to males, even in cultures where breasts are not under cloth,[17] is strong evidence that breasts are evolved sexual releasers. That this function of breasts is probably unique to hominids seems directly related to the upright stance, for only in the upright can such a development stand out visually. By contrast many of the quadruped primates, when in oestrus, display a swollen and/or brilliantly coloured sex skin as a visual attractant.

In a discussion of breasts as sexual releasers, it behoves us to note that in some advanced cultures, such as the Chinese, large breasts are considered too animal-like, and the practice of binding the chest to reduce breast size has been widespread. Of course large breasts are not animal-like; but this practice is in keeping with the general Oriental motif of muting emotion and reducing animality and animal drives. Another observation in line with this point is the Japanese disapproval of twin births because the multiple birth resembles the animal litter, and, moreover, genetically selective forces seem to flow with such cultural motifs; this is demonstrated by the fact of reduced breast size among many Oriental groups as compared with other races, and a reduced occurrence of twinning among the Japanese (approximately half the average Negro rate). Along with this last point, certain African tribes which hold twinning in high esteem have a twinning rate about ten times that of the European rate (LeVine, 1967 personal communication). In the absence of historical documents, it is probably pointless to speculate whether culture dictated genetic selection in these two instances or vice versa. It is safest to assume a feedback process between the two.

With regard to the phenomenology of courting, the advent of the " sweater girl " over the entire westernized globe has demonstrated how the breasts can be actively used as an attractant; a lazy stretch, a fully erect back, are all natural responses of seeking females. The consequent male arousal and courting responses need not be documented in any detail, but we should note that once the distance receptors are stimulated and the female is approached, the smile and the meeting of eyes become further media of drawing together. Such prolonged confrontation of the eyes in courtship usually occurs just before intimacy is completed, and while cultural differences enter at all points, the essential nature of the encounter is probably everywhere the same. That the encounter of eyes in the *en face* position has a good deal of further meaning in human relationships is clear, and it will be discussed again in the section on infant-adult attachments.

It is of interest that female inhibitions in courting are themselves attractive to males, and the girlish giggle and shyness may be viewed as ritualized attractants which indicate a woman's interest in the other sex. The recent work of Eibl-Eibesfeld and Hans Hasse (unpublished) in a wide variety of societies indicates that such behaviour (as a response to complimentary remarks by a male) are, in all probability, species-wide. As for the male, the more shy the girl is, the more she giggles and the more her face turns red, the more attracted he becomes. Thus the reddened face is a message, and during courting the blush and the askance look indicate the female's knowledge that she is being looked at desirously and, at the same time, that she is not undelighted at this attention. The male finds this attractive and, among other things, such behaviour never calls his dominance in question, for it is a ready admission of submission. In general, non-cryptic animal coloration is an adaptive attention-getting device, and, psycho-pathological blushing aside,[15] blushing appears to fulfil such an evolutionary function (see also Darwin's discussion[10]).

Inhibitions towards courting seem quite as universal as courting itself. Males, particularly young males, tend to court with an eye on the male dominance hierarchy, so to speak, and feelings of subdominance within the male hierarchy are often associated with sexual inhibitions. To take an extreme but illustrative example: among the Marquesas, thwarted adolescent love is often followed by the feeling that peers are derisively laughing at one, and this is the most frequent cause for suicide in adolescent males.[32] The Marquesas are also illustrative of a society in which adults permit adolescents fairly full sexual freedom and in which adolescent inhibitions, such as they are, are largely self-imposed and based on relative self-esteem within the peer group.[46]

Inhibition to courting is not a uniquely hominid trait, for DeVore[12] speaks of functional castration (following the Freudian castration complex) in subdominant baboon males who are at once attracted and inhibited about pursuing the female. In humans these inhibitions are played upon differently in different cultures, as are the courtship procedures, but it is probable that they everywhere reflect the male's self-image regarding his position in the dominance hierarchy; so called adolescent infertility (discussed by Montagu[39]) is probably in some degree due to such functional impotence. It thus does not seem unreasonable to speculate that functional impotence is an adaptive mechanism which tends to assure that the mating males will be the most mature and the most ready to rear a family.

In this vein, such sexual inhibition and shyness, so far as it creates conflict which is emotionally arousing, probably serves often to solidify attachments among humans; Davenport,[11] for example, has so described the function of proscriptions against premarital liaisons in a Melanesian society. For that matter, sexual inhibition seems to be present in all human societies, and unlike the lower primates [12] copulation rarely occurs in front of conspecifics and usually takes place in the dark [17]—with the exception, of course, of orgiastic ceremonies.

This shyness may well be related to the frequency with which copulation occurs in man and the fact that he is vulnerable to attack, especially by rivalrous males, while he is so engaged. Evidence for this assumption is the frequent fantasy common in various cultures, especially in sub-dominant males, that another larger male will in some way render him impotent. [43] Secretive sexuality serves also to reduce overt sexual rivalry, and to that extent maintains the integrated co-operative organization of the family and the wider group. Regarding the possible phylogeny of this behaviour, it is important to note that males of many species, particularly those which may be preyed upon, are sexually inhibited when in strange places. [17]

POLYGAMY

The opposition of polygamous drives, especially as seen in males, to the counterpull of developed attachments is another case of opposing drives ending in compromise, since each serves an adaptive function. To spell out what is almost obvious, the polygamous male ensures that all females will be fecundated, whereas the stable mating pair ensures for the young the protection and nurture afforded by a father.

For all practical purposes polygamy and polygyny are identical terms, since polyandrous societies are extremely rare, [40] and investigation always reveals very special circumstances which characterize them, such as too few women. [36] Needless to say, all societies deal with the opposed forces of polygamy and monogamy in different ways—for example, by legalized polygamy, illegal polygamy, concubinage, enforced monogamy or prostitution—depending upon the environment provided by a given culture. Legal polygamy is practised in 418 of the 554 societies rated by Murdock, [40] or in 75 per cent of the world's societies; whereas 25 per cent are characterized by monogamy.

This leads directly to a second aspect of polygamy, that

which has to do with male social status, i.e. the male's position in the dominance hierarchy. While this will be discussed later as well, it is important here to point out the fierce possessiveness males have for their women and the simultaneous importance of not being outdone or cuckolded by another male. Thus, being made cuckold is perhaps the most frequent cause for within-group murder over the entire world, and it appears that most cultures exonerate the jealous husband—the initial possessor—who was driven " temporarily insane " by his shame and hurt. The deep emotions the human male feels in this situation clearly have as much to do with loss of self-esteem as with loss of the woman's love, and loss of self-esteem is in turn dependent on the male's view of his current position *vis-à-vis* others in the social hierarchy.

In the same regard, the possession of several women is often involved with inner feelings of well-being, of being " on top of the world "; in many cultures one can witness the importance of male ' bragging-sessions ' about the number of women one has had, and in some societies the relationship between status and number of woman possessed is openly and officially acknowledged. It appears that having the fealty of several women brings out feelings of ascendancy in a man, and also earns the awe and respect of the males around him. Similar feelings and social status may also accompany the monogamous possession of a particularly beautiful or highly esteemed female.

It is very likely that the male's possessive attitude towards the female, once the claim on her has been personally made and /or publicly proclaimed, is an important factor in the various cultural rules concerning incest and exogamy. Thus, if it is not permitted that a clan member come into competition for another's female, much potential bloodshed is avoided, and numerous cultural rules have consequently arisen to prevent such rivalry.

To attribute the universality of incest taboos and exogamy to universal stages in individual development, as psychoanalytic theory demands (the desire of the male child to possess the mother and destroy the father), does not appear reasonable. As stated elsewhere, the fact that two events are related and that one comes earlier than the other does not prove that the first event caused the later one. Rivalry between males, and to a lesser extent between females, most likely has its origins in man's phylogenetic history, and it seems to be mediated in part by hormonal effects on the central nervous system. The primary (evolved) emotions of possessiveness and jealousy characteristic of either mate, and a fear of retribution by an outraged or robbed possessor or potential possessor, seem in turn to be the main ingredients which have led to exogamy and incest taboos.

SOME ASPECTS OF EARLY DEVELOPMENT

Crying, Holding and Caretaking

The very first behaviour exhibited by the human newborn is the cry. This is a common mammalian occurrence and seems to serve the general mammalian function of exciting the parent to caretaking activities. In dogs, for example, a puppy removed from the nest immediately starts to cry and continues until exhausted. The bitch will usually become extremely excited, seek the source of the cry until the puppy is found and then fetch it back. What we have here, clearly, are two complementary evolved mechanisms, neither of which has to be learned.

In the human species, similarly, we have demonstrated that within hours after birth and before the first feed most crying infants will quiet when held and carried. Consider how this cessation of crying co-ordinates beautifully with the intense anxiety felt by the parent until the infant is quieted. Apart from caretaking and feeding, body contact is the inevitable result of crying, and the human baby does as well as the macaque in getting next to the parent even though it lacks the ability to cling. There seems little doubt that such contact is normally a mutually reinforcing experience, and tactile contacts of one form or another remain an important means of relating throughout the lifespan.

Smiling

Smiling is also quite clearly an evolved mechanism.[2, 18] It is universally present in man and it has the same or similar interpersonal function everywhere, that of a positive greeting or of appeasement. Smiling is first seen in reflexive form in newborns, including prematures, when they are dozing with eyes closed, usually after a feeding. Even at these early ages, however, smiles can also be elicited by a voice or by rocking the infant and, since it occurs in infants whose gestational age is as low as seven months[19] it would not be surprising if smiling, like thumb-sucking, is eventually found to occur *in utero*. Visually elicited smiles occur somewhat later than aurally elicited ones, though they are occasionally seen within the first week of life. These are called social smiles, since they occur most readily when the eyes of infant and adult meet, but in the auditory mode too the preference for a voice over other sounds also marks such smiles as ' social '.[52]

The major function of smiling, then, from a very early age is responsivity to another. It provides an important means of attachment between adult and infant, and in later life it lends ease and promotes attachment in a wide variety of social encounters. It is also widely displayed between adults as a gesture of greeting and appeasement, and it is a major means of either precluding or overcoming dissension and angry feeling.

It is pure surmise, of course, whether the smiling response appeared phylogenetically as an adult-adult mechanism or as an adult-infant mechanism. It is most akin to the ' frightened grin ' in other primates, a gesture frequently made by a subordinate animal when passing close to a dominant one,[26] and human smiling may well have originated with such a gesture in an evolutionary ' turning to the opposite '.

Watching, Cooing, Laughter and Play

The importance of the auditory and visual receptors in the young human infant seems directly related to its general motor immaturity. Thus, the eyes begin to search for form and movement in the environment soon after birth,[14, 24] and by two weeks of age over 50 per cent of all infants will visually follow a moving person (Bayley, personal communication, 1961).

At about two months the infant's searching for the adult face can be very impressive. If held at the shoulder an infant may hold its unsteady head back to get a view of the holder's face, craning its neck like an inquisitive goose. One is left with the ineluctable feeling that seeking the *en face* position is itself an evolved mechanism. Supporting this contention are several experimental studies which find the face a preferred stimulus for most infants, including newborns [14], and disclose the fact that the adult feels ' looked at ' for the first time just before the onset of social smiling. The human orientation towards the face of another is undoubtedly bound up with many aspects of evolutionary adaptation, including the upright stance, the relative hairlessness and rich musculature of the face, and, perhaps most importantly, the high degree of interpersonal communication.

A few weeks after *en face* smiling starts, the infant begins to coo at the beholding adult who, in turn, usually feels an irresistible urge to respond, and as a result much time may be spent in such happy ' conversation '. Feedings and sleep have by then decreased, and normally more and more time is spent in direct social interactions.

A more robust order of interaction is initiated by laughter, usually at four months, when the baby and caretaker begin to engage in mutual play. The joy the adult feels in this engagement is probably no less an evolved mechanism than the laughter of the baby, and doubtless such mutually reinforcing emotion results in attachment.

The factor of time spent together also solidifies attachment, and this is served, of course, by all the mechanisms described above.

AUTONOMY AND ITS COUNTERFORCES

In those children who have formed attachments, the drive for independent action becomes insistent as motor independence is achieved, early in the second year; and foolhardy bravery, extreme negativism and possessiveness appear almost simultaneously.[4] As a consequence, parental watchfulness becomes extremely important at this age, and parental protectiveness may now be more frequently seen. It is interesting that in monkeys, such as the Japanese macaque (personal observations), the dominant male develops a ferocious protective response to the newly motile young; and, in fact, since these monkeys travel on the ground as a troop, it becomes largely his job to protect against possible predation. Similarly, while protective responses on the part of human adults may be experienced as purely volitional acts, the intense emotions and vigorous activity aroused when a child is endangered seem, instead, reflexive in nature.

Besides the protection provided by a parent, toddlers have numerous phylogenetically adaptive responses which offset the dangers contingent on investigativeness and motoric bravery. The fear of falling from heights is quite clearly ' built into ' most animals, including the human ; [21] further, no matter how brave the toddler, he usually melts into loud sobs when he finally realizes he is lost. Also, various fears become characteristic of most children; fear of large animals and kidnappers, fear of being lost, nightmares of being captured and eaten, appear not to have exclusive origin in parental warnings, although they may become exacerbated by them.[30] Thus, constant readiness for danger seems to typify man as it does all animals, particularly those which may be preyed upon.

BEHAVIOUR WITHIN GROUPS

Dominance-submission Hierarchy

All primates show various forms of social interactions and there can be little doubt that man's social interactions are also based on related evolved mechanisms. One outstanding similarity among almost all highly socialized vertebrates involves the establishment of a dynamically stable dominance hierarchy. Pertinent animal data have already been discussed, so that it now remains to describe hominid hierarchies.

Ascendancy and submission are everywhere present in the institutions, concepts and activities of man. Osgood,[41] for example, has demonstrated that an extremely high percentage of the emotional words of many languages in many culture areas involves references to relationships of power ascendancy over, or submission to, others.

As was mentioned previously, females often achieve status vicariously through their mate and their children, and they do not necessarily feel this to be second-best achievement; it seems, by and large, more characteristic of them not to vie directly for position in the social hierarchy. Instead, as in baboon groups[12] the female vies with other females in terms of position of the males associated with her (for example, husband and son) in the dominance hierarchy.

Many familiar behaviours appear to stem from the hierarchical arrangement of groups. Looking at others and being looked at are very important in hominid life and seem to reflect the importance of assessing oneself in relation to others in the group. It is likely that clothing, adornment, self-grooming— in short, the evolution of ' good looks '—are associated with this aspect of hominid adaptation.

Along with the constant looking at others and assessment of their externals, humans have a correlated interest in what others are thinking, that is, the assessment of their thoughts. With this tendency goes a sensitivity to public derision, familiarly experienced as the tendency to see derision in what may be the innocent laughter of a group of strangers. To call this process the projection of one's own derisive feelings, as parlour psychoanalysts do, is to neglect the non-pathological, socially adaptive aspect of such behaviour. Clearly, to be socially shamed is a powerful deterrent against group-disrupting activity.

Finally, it should be emphasized that wherever the dominance-submission hierarchy is found it serves to prevent excessive conflicts in that it offers a way of stabilizing a group. The moments of active challenge are highly dramatic and therefore often receive emphasis, but by and large viable groups are those which achieve substantial stability and co-operation within a mutually accepted hierarchy, where divisive tendencies are balanced by integrative ones.

One-up versus *one-down*

The subjective aspect of interpersonal encounters among hominids often involves feelings of being one-up or one-down, and these feelings seem to stem from the hierarchically oriented nature of the species. To give a common example, in either purchasing or selling, avoiding the one-down position and attaining the one-up position is characteristically more important than the amount of money gained or lost; the proof is that often the very same subjective feelings (triumph versus hurt and anger) follow either petty or substantial gains or losses. Psychological sub-theories are often based on this fact, and Festinger's [16] ' dissonance ' theory is based on the observation that persons attempt to justify acts already completed, that is, they attempt to make sure they are one-up or else attempt to reverse one-down situations. For example, one often window-shops *after* making a purchase for reassurance that it was a ' good buy. ' Berne's [5] descriptions of ' games ' people play with each other also tend to be examples of persons vying for a one-up position. In the same vein, Haley [25] has graphically described the psychotherapeutic treatment of a neurotic as the vying between therapist and patient for the one-up position. Haley wryly concludes that when the patient finally realizes he cannot become one-up on the therapist, because of the very nature of the doctor-patient relationship, it is called a ' cure '.

The same emotions seem to characterize negotiations between groups. In collective bargaining between unions and employers, or in tariff discussions between nations, and so on, it is of considerable importance that each side leave the bargaining table with the feeling that it has gained something; if there is the feeling by one party that it has not gained, or gained less than the other party, hostility usually again erupts within a short period of time. [44]

Thus the sense of being one-down and the consequent attempt at realigning the position of power (one-upmanship) is a driving force in many hominid activities. So far as I can see, there are no facts which support the notion that this competitive relationship with other males and other groups ever dissolves into pure amity; and, in fact, amity without enmity is found in no primate group. [3, 12]

Displaced Aggression and Appeasement Behaviour

Much of man's life involves linguistic contacts, and many activities which appear as motor acts in lower species appear linguistically in man. For example, language usually plays a substantial role in aggression, courtship and greeting behaviour, and the specific vocabulary is incidental to the motivation underlying these behaviours. Similarly, *displaced aggression* is almost

E

always verbal in man, and we usually call such behaviour gossip, grumbling or scandal.

To give but one familiar example: in an organized hierarchy, such as an army or a police force, an official system of deference is in force that gives naturally occurring enmities the outward appearance of smooth operation and turns within-group aggression into gossip, grumbling and scandal. In addition to gossip and other forms of displaced aggression, most human groups depend on greeting and appeasement ceremonies to maintain social bonds even as analogous ceremonies are necessary for viable social functioning in other species.[34]

Goodall[23], for example, has observed greeting and appeasement behaviour among free-living chimpanzees and has been impressed with their 'human' qualities. She has described how there is tension within a group when a dominant male enters it, until he shows to the group he is not in an aggressive mood. He may show this by touching others on the body or hands, or by nuzzling; whereupon the group visibly relaxes.

Greeting ceremonies in man are, needless to say, quite similar. The touching or raising of open hands is a frequent form of greeting, as is the kiss, an embrace and submissive lowering of the head. The display of the smile can be perceived over a considerable distance and is therefore particularly effective in reducing tension. Relaxed posture, of course, is also important, and a stalking posture is readily perceived. Words of greeting are equally important, and a threat versus a friendly greeting are usually distinguishable even if the language is not known.

Like other animals, man can be deceptive, and his greeting ceremonies are sometimes used for deception. Most often, however, they become ritualized as an everyday lubricant in social relationships. Consider, for example, a public argument between two males, or an athletic match such as tennis, and the importance attached to the two contestants' exchanging a few friendly words afterwards to overcome the engendered tension. If mutual appeasement does not occur following an aggressive encounter, tension continues and anger builds up. It is clear that in man a smile, a touch of hands or a few ritualized words can do wonders in dissipating growing anger.

Many cultures, as in France or Japan, institutionalize smiling, touching of hands, kissing, embracing or bowing as parting and greeting ceremonies. It is significant that day-to-day relationships go more smoothly in those countries than in a young immigrant country such as the United States, where there is still considerable awkwardness about greetings, appeasement gestures and ceremony. Although various appropriate gestures are used in the United States, they are not yet nationally institutionalized; for example: one person may offer his hand for a handshake whereas the other, not expecting this, has his hands in his pockets. It is not surprising, then, that a fairly constant anxiety surrounds interpersonal relations in this country, and newspaper columns on social etiquette are, probably as a consequence, extremely popular.

To demonstrate that greetings do indeed perform an important function, one can perform the experiment of not smiling or saying " Hello " to those whom one normally greets. Before many mornings are past the level of aggression will have risen considerably, and yet no direct act of anger will have been perceived. Contrariwise, the institutionalization of greeting and appeasement gestures, where none existed before, may lighten the social atmosphere considerably. For example, a chronically unhappy group can be made more spontaneous and warm if a person high in the hierarchy initiates greetings, smiles frequently, engages in interested conversation, and thereby precludes a preoccupation with feelings of anger. Social gatherings may have a similar function for groups; parties, dances and ceremonies of various sorts can serve, among other things, to reduce the build-up of antagonisms.

Averting one's eyes is usually a gesture of appeasement, and it is often necessary to keep aggression low, especially between strange males, since the direct confrontation of eyes can be taken as a challenge to one's dominance. This is an interesting phenomenon which holds over a wide range of species, and one can, for example, elicit challenge after challenge in a zoo by staring into the eyes of some dominant animals, ranging from birds to primates; subordinate animals, by contrast, merely avert or shut their eyes.

Averting of the eyes and head is closely related to turning

the back and presenting the rear as a gesture of appeasement among primates; in a group whose members are familiar with each other such motor patterns usually serve to prevent further attack. Interestingly enough, very similar behaviour is to be found on any schoolground during recess, particularly among the boys, as when one calls off the ' attack ' of another by turning away and crouching. The message transmitted seems to be : " I'm in no mood or not afraid enough to run from you ; but I acknowledge your greater potency ".

The present essay has given a cursory sketch of man's social behaviour drawn within an evolutionary framework. I have omitted discussion of many important aspects of human social behaviour which seem to be present in all societies and which are thus species-specific: friendship, religion and mourning, dance, song, festivity and humour, among many. All share the common importance of binding members of the group to one another and, as a matter of fact, functional interpretations of some of these behaviours are available. For examples, see Wallace[50] on religion; Freud[20] on humour; Evans-Pritchard[13] on dance.

It seems pointless to carry on a debate about whether these and the other behaviours we have discussed are, or are not, products of phylogenetic adaptation. Only the accumulation of properly designed empirical work can settle such issues, and in our laboratories we are making a start in this direction. For example, it has been proposed that the cry of a baby is phylogenetically adaptive in that it elicits caretaking reactions in human adults. As an initial study we have made psychophysiological measures of responsivity to a series of sounds matched in loudness and complexity, including baby cries. As predicted, the baby cries drew a greater level of response in all adults, and this was most pronounced when the subjects were mothers.

In general, behavioural experimentation on human subjects performed within an evolutionary framework is practically non-existent, despite the great possibilities, because psychologists have simply not thought in this way.

In addition to the need for empiricism, we should not lose

sight of the basic logical principle which guides an evolutionary approach to man. Stated broadly, it asserts that everything which man does must, at some level, reflect his biological make-up and his evolutionary past, and that man's cultural and biological nature are two aspects of the same macro-feedback system. Thus, although it is true that societies and civilizations change with time, the conservatism of basic institutions and behaviour is necessarily matched by the conservatism of man's genotype—for they are indeed parts of the same phenomenon, the one we call man.

Despite obvious problems that beset this broad approach, progress in the science of man demands an initial overall conceptualization which can serve as a reasonable guide to detailed exploration. The other alternatives (for example, that of the reflexologists) have stressed the search for basic units which, it was hoped, would by accretion lead to a scientifically based conception of man; the history of biology and psychology, however, has proved the extreme limitations of such an atomistic approach.[22]

Finally, it is my hope that the experimental examination of corollaries stemming from the evolutionary point of view will soon begin to appear in scientific journals which concern themselves with the behaviour of man, even as they have appeared for many years now in journals on animal behaviour.

REFERENCES

1. AINSWORTH, M. D. 1962. *The Effects of Maternal Deprivation: A review of findings and controversy in the context of research strategy.* Geneva. WHO. Public Health Papers, No. 14.
2. AMBROSE, J. A. 1960. The smiling and related responses in early human infancy : an experimental and theoretical study of their course and significance. Ph.D. dissertation, University of London.
3. ARDREY, R. 1966. *The Territorial Imperative: A Personal Inquiry into the Animal Origins of Property and Nations.* New York. Atheneum.
4. AUSUBEL, D. P. 1958. *Theory and Problems of Child Development.* New York. Grune and Stratton.
5. BERNE, E. 1964. *Games People Play: The Psychology of Human Relationships.* New York. Grove Press.
6. BOCK, R. D. and VANDENBERG, S. G. 1966. Components of heritable

60   INFLUENCES ON BEHAVIOUR

variation in mental test scores. Read at Second Louisville Conference on Human Behavior Genetics.

7. BOOTH, P. B. 1966. Sex differences in manifest content of South
African dreams. M.A. thesis, University of Chicago.
8. BUYTENDIJK, F. J. J. 1962. The phenomenological approach to the
problem of feelings and emotions. In *Psychoanalysis and Existential
Philosophy*. Ed. H. Ruitenbeek. New York. Dutton.
9. DARLINGTON, C. D. 1958. *The Evolution of Genetic Systems*. (2nd ed.).
Cambridge. University Press.
10. DARWIN, C. 1873. *The Expression of the Emotions in Man and Animals*.
London. Murray.
11. DAVENPORT, W. 1965. Sexual patterns and their regulation in a
society of the Southwest Pacific. In *Sex and Behavior*. Ed. F. A.
Beach. New York. Wiley.
12. DEVORE, I. (Ed.). 1965. *Primate Behavior: Field Studies of Monkeys
and Apes*. New York. Holt, Rinehart and Winston.
13. EVANS-PRITCHARD, E. E. 1965. The dance. In *The Position of Women
in Primitive Societies and Other Essays in Social Anthropology*. New York.
Free Press.
14. FANTZ, R. L. 1966. The origin of form perception. In *Frontiers of
Psychological Research: Readings from Scientific American*. Ed. S. Cooper-
smith. San Francisco. Freeman.
15. FELDMAN, S. 1962. Blushing, fear of blushing, and shame. *J. Am.
psychoanal. Ass.* **10,** 368.
16. FESTINGER, L. 1966. Cognitive dissonance. In *Frontiers of Psychological Research: Readings from Scientific American*. Ed. S. Cooper-
smith. San Francisco. Freeman.
17. FORD, C. S. and BEACH, F. A. 1952. *Patterns of Sexual Behavior*. New
York. Harper.
18. FREEDMAN, D. G. 1964. Smiling in blind infants and the issue of
innate *vs.* acquired. *J. Child Psychol. Psychiat.* **5,** 171–184.
19. FREEDMAN, D. G. 1966. An Evolutionary Framework for Behavioral
Research. Presented to the Second Conference on Human Behavior
Genetics, Louisville.
20. FREUD, S. 1938. Wit and its Relation to the Unconscious. In *The
Basic Writings of Sigmund Freud*. Trans. A. A. Brill. New York.
Modern Library.
21. GIBSON, E. R. and WALK, R. D. 1960 (April). The visual cliff.
Scient. Am. **202,** 64.
22. GOLDSTEIN, KURT. 1939. *The Organism*. New York. American
Book Co.
23. GOODALL, J. 1965. Chimpanzees of the Gombe Stream Reserve. In
Primate Behavior: Field Studies of Monkeys and Apes. Ed. I. DeVore.
New York. Holt, Rinehart and Winston.
24. GREENMAN, G. W. 1963. Visual behavior of newborn infants. In
Modern Perspectives in Child Development. Ed. A. J. Solnit and S. A.
Provence. New York. International Universities Press.

25. HALEY, J. 1958, Spring. The art of psychoanalysis. *Etc.: A Review of General Semantics*, **15,** 190.
26. HALL, K. R. L. and DEVORE, I. 1965. Baboon social behavior. In *Primate Behavior: Field Studies of Monkeys and Apes.* Ed. I. DeVore. New York. Holt, Rinehart and Winston.
27. HARLOW, HARRY F. and HARLOW, M. 1966. Learning to love. *Am. Scient.* **54,** 244.
28. HARRIS, G. 1964. Sex hormones, brain development, and brain function. *Endocrin.* **75,** 627.
29. HUXLEY, J. S. 1958. Cultural process and evolution. In *Behavior and Evolution.* Ed. A. Roe and G. G. Simpson. New Haven. Yale University Press.
30. JERSILD, A. T. 1954. Emotional development. In *Manual of Child Psychology.* Ed. L. Carmichael. New York. Wiley.
31. KINSEY, A. C. 1953. *Sexual Behavior in the Human Female.* Philadelphia. W. B. Saunders.
32. LINTON, R. 1939. Marquesan culture. In *The Individual and His Society: the Psychodynamics of Primitive Social Organization.* Ed. Kardiner. New York. Columbia University Press.
33. LORENZ, K. 1965. *Evolution and Modification of Behavior.* Chicago. University Press.
34. LORENZ, K. 1966. *On Aggression.* New York. Harcourt, Brace and World.
35. MACCOBY, E. (Ed.). 1966. *The Development of Sex Differences.* Stanford, Cal. Stanford University Press.
36. MALINOWSKI, B. 1956. In *Marriage: Past and Present.* Ed. M. A. Montagu. Boston. Sargent.
37. MASTERS, W. H. and JOHNSON, VIRGINIA E. 1965. The sexual response cycles of the human male and female : comparative anatomy and physiology. In *Sex and Behavior.* Ed. F. A. Beach. New York. Wiley.
38. MEAD, M. 1939. *From the South Seas: Studies of Adolescence and Sex in Primitive Societies.* New York. Morrow.
39. MONTAGU, M. A. 1946. *Adolescent Sterility: A study in the comparative physiology of the infecundity of the adolescent organism in mammals and man.* Springfield, Ill. C. C. Thomas.
40. MURDOCK, G. P. 1957. World ethnographic sample. *Am. Anthrop.* **59,** 664.
41. OSGOOD, C. E., MIRON, M. S. and ARCHER, W. K. 1963. The cross-cultural generality of effective meaning systems : *Prog. Rep. Cent. Comp. Psycholing.* University of Illinois.
42. PIAGET, J. 1950. *The Moral Judgement of the Child.* Glencoe, Ill. Free Press.
43. ROHEIM, G. 1950. *Psychoanalysis and Anthropology: Culture, Personality and the Unconscious.* New York. International Universities Press.
44. SAWYER, J. and GUETZKOW, H. 1965. Bargaining and negotiation in international relations. In *International Behavior: A Social-Psychological*

Analysis. Ed. H. C. Kelman. New York. Holt, Rinehart and Winston.

45. SPIETH, H. T. 1952. Mating behavior within the genus Drosophila (Diptera). *Bull. Am. Mus. nat. Hist.* **99,** 395.

46. SUGGS, R. C. 1966. *Marquesan Sexual Behavior.* New York. Harcourt, Brace and World.

47. TANNER, J. M. 1961. *Education and Physical Growth.* London. University of London Press.

47a. TINBERGEN, N. 1953. *The Herring Gull's World: A Study of the Social Behaviour of Birds.* London. Collins.

48. UEXKÜLL, J. J., VON. 1926. *Theoretical Biology.* New York. Harcourt, Brace.

49. WADDINGTON, C. H. 1960. *The Ethical Animal.* New York. Atheneum.

50. WALLACE, A. F. C. 1966. *Religion; an Anthropological View.* New York. Random House.

51. WHITE, L. A. 1949. Energy and the evolution of culture. In *The Science of Culture.* 363–393.

52. WOLFF, P. H. 1963. Observations on the early development of smiling. In *Determinants of Infant Behavior,* II. Ed. B. M. Foss. London. Methuen.

53. WYNNE-EDWARDS, V. C. 1962. *Animal Dispersion in Relation to Social Behavior.* New York. Hafner.

54. YOUNG, W. C. 1965. The organization of sexual behavior by hormonal action during the prenatal and larval periods in vertebrates. In *Sex and Behavior.* Ed. F. Beach. New York. Wiley.

ENVIRONMENTAL INFLUENCES

Chairman: Professor Sir Alan Parkes

ENVIRONMENTAL CONTROL OVER FOUR MAJOR PATHS OF MAMMALIAN EVOLUTION

JOHN B. CALHOUN

Unit for Research on Behavioral Systems, Laboratory of Psychology, National Institute of Mental Health, Bethesda, Maryland 20014 USA

INTRODUCTION

THE present rate of population increase and the rapidity of cultural and technological change represent a unique phase of human evolution. Resulting challenges and crises provide both opportunities and dangers; by intent or by chance one or other of the several paths open to the human future will be taken. In fact, tendencies towards embarking on each of four major paths are now evident. Three of these are retrogressive, in the sense that each makes more difficult the maintenance of a culture as complex as that of the present. They are :

1. The tendency for each individual to maximize his isolation from associates and from any type of environmental change ;

2. The tendency to maximize gratification at the expense of capacity to deal with diversity ;

3. The tendency to minimize physical distance from associates while at the same time blocking out perceptual awareness of them or of surrounding environmental conditions.

In contrast to these retrogressive tendencies there is also a fourth path—a tendency towards enhancing capacities to cope with complexity and diversity. This path represents a progressive pattern of evolution in the sense that it increases the range of future options; it broadens the base of adaptability.

If man is to have a say as to which of these paths he will prefer, he will need a more precise insight into the processes involved. He cannot withhold judgment until observations on his own course alone provide the basis for confirming the wisdom of his choice. Every portion of a course pursued limits

the opportunity for taking alternate courses. Means must therefore be sought for making the most judicious choice as rapidly
as possible.

Simulation of the forces and processes affecting man's present
course may be instituted by establishing and studying large-
scale social systems of mammals other than man. Even with
short-lived mammals which attain full sexual maturity in less
than a year, such studies will require a minimum eight- to
fifteen-year duration. They will be large scale and relatively
costly as compared with any experimental studies so far executed
with mammals relating to population dynamics, evolution and
behaviour. In our laboratory we have for several years discussed
the objectives of such studies, their theoretical foundation and
the methodological problems to be surmounted for their im-
plementation. Much of our past and projected research programmes bears directly on these issues. However, my intention
here is not to marshal a body of relevant ' hard ' data, but
rather to present a general impression of the present stage of
our thoughts about this potentially useful field of investigation.

Every species of mammal consists of several to many populations, each inhabiting some finite geographical area which at
any moment in time is separated from other such areas by varying degrees of ecological or physical barriers. Furthermore, the
members of any given population are usually unequally distributed over the area available to them, even when their habitat
is relatively uniform. Instead, the social bonds which develop
among individuals hold groups of varying size together. Thus,
in carrying out experimental population research we are faced
with two primary issues. First, we seek to determine what is
the minimal area required for a population in the sense that it
will permit an interaction between the genetically determined
capacities of the contained individuals and the restraints or
opportunities for their expression imposed by the physical structure of the environment. Second, there is the problem of
determining the optimum group size for any particular species
to be studied.

I have elsewhere [5] explored in detail considerable empirical
data about the social use of space and its theoretical implications.
It will be helpful to review some of the major conclusions and

hypotheses of this study as a background for the present discussion of the experimental study of those four major paths of evolution which may be taken by man in the future.

Even the most social species of mammals derive their evolutionary origin from forms ecologically comparable to many existing species of rodents, in which each individual leads a relatively solitary way of life interrupted by occasional contacts with similar neighbours. Each such individual establishes a single or several nearby places to which it returns for periods of rest. From such ' home points ' or home range centres it makes excursions out into the surrounding habitat. Progressively fewer trips terminate at increasing distances from the home point. This means that as the distance from its home increases, each individual makes less and less effective use of its environment. Unutilized resources cannot long be tolerated in nature. As a population, such species maximize the utilization of resources by filling up space with additional individuals until adjoining ranges overlap sufficiently to produce a uniform impact on every point in the area inhabited by the population.

When a population attains this equilibrium, the distance between the homes of neighbours becomes approximately two-thirds of the radius of the range of any individual. Due to this degree of overlap of range, any given individual will often contact its six nearest neighbours and, less frequently, its twelve next-nearest neighbours also. These contacts lead to social bonds which produce an attraction between individuals which partially offsets the repulsion between them required to produce the spacing which leads to a uniform use of the environment. As a result, a loose clumping develops in which the home sites shift slightly towards each other. These clumps, which I have designated as ' constellations ', will contain between seven and nineteen individuals, with an average of twelve.

With the passage of evolutionary time, these dispersed groups became more compact until all members were residing at the same place, at least in the sense of being quite near each other during periods of rest. Some species also evolved the practice of moving through space together as a pack rather than simply returning to the same site after individual foraging trips.

Where such a relatively constant group size comes to characterize a species, this number of individuals may be said to be its optimum or basic number, N_b. Numerous examples where N_b approximates 12 adult individuals may be found in most orders of mammals including the primates. Data from culturally primitive gathering peoples strongly suggest that man emerged through such a lineage of species having N_b = 12 adults. I further suspect that modern technological man still bears the yoke or wears the diadem of an N_b = 12.

Although I believe that an N_b represents the central tendency for the evolution of social group formation in mammals, many species can be found which customarily, or at least often and for considerable periods of time, maintain much larger aggregations. These larger aggregations become the basic number, N_b, for each such species. Wherever an N_b significantly greater than 12 adults has come to characterize a species, there appears to have been a single type of influencing environmental factor in the evolutionary history and continued existence of the species. This factor is that there is, and has been, a marked spatial restriction of some needed resource. Beaches appropriate for mating and whelping of seals, water holes for ungulates and caves as resting sites for bats represent typical examples. Even in such markedly aggregating species, the degree of aggregation often seems greater than is necessary for compatibility with the observed availability of unused sites. Here the general explanation appears to be compatible with a psychological process I have observed [2] in my experimental studies of the social behaviour and population dynamics of domesticated strains of Norway rats.

Translating this specific case into general terms, we may state the process and the circumstances in which it takes place as follows. The ranges of individuals or of groups overlap. Within the range of each individual there are a few localized resources or places appropriate to a particular behaviour. Sometimes an individual will exhibit the behaviour while it is alone at one of these places, other times and purely by chance, one or more others will arrive at the same place and simultaneously engage in the appropriate behaviour. With increasing frequency of such contiguity each individual becomes a secondary reinforcing stimulus to the others present and there engaging in the behaviour appropriate to that place. Gradually, that assembly of stimuli which designates a place as appropriate for initiating the particular behaviour comes to include the presence of other individuals. Among the several places of otherwise similar characteristics, the animals will most frequently seek out those which include this acquired definition of appropriateness. This process of biased visitation may continue to the extent that within any one individual's range it will visit only one of all the prior appropriate places for engaging in the behaviour in question; furthermore, many individuals will shift their homes toward this single place. I have designated this process of development of excessive aggregation as the ' behavioural sink ' phenomenon, in analogy to geomorphic sinks. Geomorphic sinks without adequate drainage become centres of stagnation from decaying organic matter; behavioural sinks become centres of social stagnation resulting from the consequences of excessive contact which block the expression of the normal behavioural repertoire. Where the members of a species are repeatedly exposed to conditions fostering aggregations greater than that to which they have previously been adapted, they must undergo evolutionary changes compatible with continuing life under local crowded conditions. Understanding what the basic alteration must be requires examination of the major consequence of social interaction.

Initiation of any social interaction requires the meeting of at least two individuals. Persistence of a behaviour involved in such interactions for a sufficient length of time almost invariably produces a refractory state, during which the individual will no longer engage in the appropriate behaviour, even though the individual may meet another which is in a need state for engaging in that particular behaviour. When such an encounter develops between one individual in a refractory state and another in a need state, the inappropriate behaviour of the former also produces a refractory period in the latter which may be designated as ' frustrating ' since its need state at the time of the encounter was inadequately satisfied. A gratifying refractory period results only when each of the individuals in an encounter are in comparable need states. Each individual will attempt to maximize the amount of time it spends in gratifying refractory states.

Where a species has developed a basic N_b group size and is able to maintain it approximately, some average rate of contact by each individual with its associates will develop. In order to maximize gratification, each individual must develop a particular level of intensity or duration of interaction since such intensity or duration of interaction determines the duration of the refractory period. What is aimed at in the overall social system is a situation in which when one individual emerges from a refractory state—that is, it enters a need state—it will have the highest probability of meeting another in a similar need state. Attainment of this objective has an interesting consequence. Attainment of maximal gratification necessitates experiencing an equal amount of frustration. Since this is so, it follows that the optimum state is one in which gratification is exactly balanced by frustration; and for this optimum state to be attained, each individual must express an intensity or duration of behaviour toward others compatible with the number of contacts per unit time characterizing the basic group size of that species. We may then expect that the physiology of such species will be adapted to such a maximization of gratification and the experiencing of an equivalent degree of frustration. Any degree of frustration up to that level may be considered as desirably stimulating. Stress characterizes frustration only when it exceeds in amount the maximal degree of gratification that is possible in such a stochastic social system.

When the actual group size declines below its evolutionarily determined basic group size, the average individual will experience less than desirable amounts of both gratification and frustration. In part, behavioural plasticity permits avoiding the suffering contingent to life in a less than optimum group size, but only to the extent that each member of the group can offset the consequences of reduced group size by increasing the intensity, or the duration, of its interactions with others. When the actual group size persists at greater than its basic group size, the average individual will experience less than the maximum amount of gratification but more than the optimum amount of frustration. Thus to the degree that the actual group size exceeds the optimum, the contained individuals will suffer from loss of gratification as well as stress from excessive frustration.

In a prior paper [7] I attempted to broaden the concept of aggression to include any act of one individual which culminated in the frustration of another. During a conversation with Professor Robert Hinde on September 5th, 1967, he admonished against this attempt; in his opinion the term ' aggression ' should be retained to apply only to those categories of behaviour which have classically been associated with this word. Accepting the wisdom of this advice, I wish now to suggest that ' repress ' and ' repression' be the generic terms—including the more limited and specific ' aggress ' and ' aggression '—to designate the processes and kinds of behaviours that lead to any type of frustration, whereby the state of frustration arises instead of one of gratification. Repression in this sense harmonizes with the psychoanalytic concept of repression as a blocking out of one's consciousness of all precepts pertaining to undesirable experiences, since the psychoanalytic concept of repression denotes a consequence of repression in its molar form of aversive acts leading to a state of frustration.

This leads us directly to the consideration of some important consequences of continuing experience with membership in a group whose numbers persist at a level considerably above that of the established basic group size. Whenever an individual enters an interaction situation by exhibiting behaviour which prior learned or evolutionary experience has dictated as appropriate, but finds that the other individual with whom it is interacting does not respond in a fashion compatible with his overtures, it not only becomes frustrated but—equally important —his apparently appropriate behaviour is not rewarded or reinforced. Thus, following each instance of frustration, the behaviour on the next encounter is likely to become somewhat deviant, with the result that, even though this next encounter is with another in a similar need state, its deviant behaviour will more likely elicit a comparable amount of repressive actions from its associate. The experiencing of undue frustration enhances the likelihood of later interactions also being frustrating.

Two processes of accommodation introduce themselves at

F

this point. First, to the degree that an individual experiences frustration above the optimal level, it begins to shrink from entering those places where it is more likely to encounter associates, and it similarly avoids those times of activity when his associates are more active. Second, to the degree that an individual experiences frustration above the optimal level, it begins to block out of its awareness surrounding stimuli which impinge upon it. In fact, we have here being established the process denoted by the psychoanalytic concept of repression, but it becomes more comprehensive in that perceptual blocking is generalized to include stimuli in general, and not just those associated with those behaviours of associates which produced the state of frustration. These two consequences of life in a greater than optimum-sized group are superimposed upon the strategy of reducing the intensity or duration of behaviour which partially compensates for increased group size.

All the remarks made above about the development of deviant behaviour, reduction of activity by physical withdrawal, and blocking of perception by psychological withdrawal, apply also to the members of a group consisting of the optimal number which forms the basic group size compatible with the evolutionary history of the species. In such an optimal group, one finds a full range of types, from the socially active—with keen perception of surrounding stimuli and with expression of behaviours most appropriate to the situations which it enters— to (at the opposite extreme) the individual which has so withdrawn both physically and psychologically that it neither initiates interactions with its associates any longer nor elicits such actions from them. In the best of all possible worlds, we may expect to find such a broad spectrum of behavioural types; however, to the extent that an existing group size exceeds the optimum, those more normal, socially active types begin to disappear and all members acquire more of the deviant behaviour and of the withdrawal characteristic of the lower end of this spectrum. For all practical purposes, when the group size attains the square of the basic group size, all individuals, despite their increased proximity, will have become so withdrawn as to have become essentially out of contact with each other in the social, behavioural sense.

There remains one final consideration before we are in a position to examine experimental approaches to behavioural evolution. Here we must rely primarily on natural history data, although experimental and observational studies of social and psychological deprivation do contribute to our understanding. I have spoken of forces leading to a basic group of 12 adults or to aggregations much larger than this. However, there are a number of species of mammals in which each individual leads an extremely solitary life, in which contacts with others are essentially limited to the minimum necessary for successful copulation and the rearing of young to weaning. On every continent the most typical of these forms are the rodents or insectivores, which have developed a nearly completely subterranean way of life. Maintaining their extreme degree of social isolation is facilitated by well developed and readily elicited aggressive behaviour, as well as behaviours increasing the likelihood of one individual avoiding another. When the marked degree of social isolation characterizing such species is considered in conjunction with the stability and simplicity of their subterranean habitat, devoid of the myriad fluctuating sounds and sights of the terrestrial world above, it is obvious that such species have largely cut themselves off from ever embarking upon an evolutionary path that would promote the origin of prototypes for species engaging in complex social relations.

In the above paragraphs, I have attempted to outline some basic concepts pertinent to developing plans for controlled studies of social and behavioural evolution. Despite the logic and empirical data I have marshalled [2, 4, 7] to support these views, I will be the first to admit that they must be considered merely as preliminary hypotheses. However, if we are to begin approaching readiness for initiating the large-scale, long-range experimental studies required to gain a more thorough understanding of the evolution of social behaviour and social organization, reliance in our planning must be placed on such formulations.

SPECIES FOR STUDY

Rapid selection to produce a form adapted to a contrived experimental habitat requires broad genetic diversity within the stock initiating a population in a spatially closed environment. We are therefore primarily restricted to consideration of species for which selection by man has already produced many geno-types. Among these we can nearly immediately eliminate such forms as dogs and rabbits, as well as the larger domesticated animals, because their more extensive spatial requirements and longer generation time would greatly increase the cost of studies. In the last analysis we are both for practical and scientific reasons limited to the house mouse, *Mus musculus*, and the Norway rat, *Rattus norvegicus*. Their attainment of prime adulthood in con-siderably less than a year, the large gene pool for each species in both laboratory strains and among widely distributed natural populations, as well as the great wealth of data about their heredity, behaviour, anatomy, physiology and ecology, make them prime targets for initial studies on behavioural evolution. The apparently simpler social structure of the house mouse and the much greater knowledge of its heredity would argue for its selection as the most appropriate subject. However, to the extent that such experimental studies of behavioural evolution may be motivated with the hope that results from them might help us in our thinking about the courses man will take by either intent or default in the relatively near future, the Norway rat holds much greater promise. I say this because for both man and rat the basic group size appears to be one of 12 adults. We must not be misled by the fact that modern man inhabits villages, cities and conurbations containing many more indi-viduals. He does so only by virtue of the fact that his culture enables him to shift rather rapidly from participation in one group to another, but with the overall effect of providing him with the number of meaningful contacts per day that character-ized simpler stages of his evolution thousands or hundreds of thousands of years ago. I shall, therefore, from my experience[1,3] with wild and domesticated strains of Norway rats, and the general insights into forces altering the degree of sociality, con-sider the construction of habitats for this species with reference to how differently composed environments may alter sociality in terms of the concepts outlined earlier in the paper.

AN ENVIRONMENTAL UNIVERSE

When we examine populations of rats in their native habitats or in extensive simulated ones[3], we find that each local group of approximately twelve individuals inhabits a discrete burrow system. One or more trails connect neighbouring burrows. These trails result from the frequent movements of rats from one burrow to and past those of its neighbours. Sometimes partial barriers, penetrated by these trails, intervene between neighbouring burrows. Even though no actual barriers may exist between burrows, the actual lack of usage of much of the intervening space between them produces a situation which, in terms of customary activities, appears as though there were partial barriers present. If one makes a map of an area containing several burrows and draws lines at right angles to the centre points of the straight lines connecting each pair of neighbouring burrows, a pattern of irregular polygons, roughly hexagonal in shape, appears. These polygons may be considered as ' cells '. Access between cells is via the ' passages ' formed by trails crossing the sides of the polygons. When any such natural system of burrows is followed, one reaches a point where trails from peripheral burrows fail to make contact with any other burrows farther out. Then the contained cells may be considered as forming a ' universe ' delimiting the extent of a particular population.

The first decision about a synthetic environment concerns the size and shape of cells. Several considerations, including the above, suggest a hexagonal shape as most desirable. Where these cells are eight feet on a side, where provision is made for both vertical and horizontal structuring of the walled cells, where allowance is made for the possibility of bridges between adjoining cells, we arrive at a minimal space for cells that can simulate the opportunity for a degree of activity similar to that achieved within cells of a natural environment by its contained members.

Unfortunately, we lack adequate information on the extent

of natural universes or the degree of genetic interchange between them. Therefore, decisions on the number of cells in a contrived rat evolution universe must be based on other sources of judgment. It should certainly be large enough for genetic drift not to cancel out the effects of natural selection. Up to the present, no detailed theoretical explorations have been made to determine the optimum-sized population. I will only add that the ecological evidence from an island to which I have taken breeding stock of wild rats suggests that the population on it is periodically reduced to as few as 150 adults. In my laboratory I maintained the core breeding stock at six cages. Each sucessive generation was formed by leaving two young females in their home cage and moving one of their male sibs to the next higher numbered cage, or in the case of Cage No. 6, to Cage No. 1. By the sixth generation of such a breeding schedule, many individuals were characterized by large white flashes on the chest, big enough almost to cover the ventral aspect. Such flashes, even small ones, were never observed on the wild-caught rats from this island. I, therefore, suspect that considerable genetic variability can be retained in a population periodically reduced to 150 adults.

For reasons to be detailed later, it would be desirable to have the numbers of individuals within cells vary among the cells of the universe. How this objective can be obtained may best be appreciated by examining Figure 1. Here is a 37-cell universe. Lines connecting the shaded cells indicate bridges or avenues of communication between cells. Numbers within cells indicate the number of other cells to which any particular cell is connected. On the basis of empirical data [7] it appears that the steady-state number of animals in any cell will be proportional to the number of other cells to which it is connected. If each cell started with twelve individuals, the total universe would contain 444 colonizing rats, but the shifting of residence produced solely by the effects of number of bridges to adjoining cells would result in a redistribution of the population to a point where the 7 central cells would each contain about twenty rats, and five groups (six in each group) of the remaining cells would contain approximately seventeen, thirteen, ten, seven and three each respectively. This process affecting distribution primarily applies to sexually immature individuals. Aggressive acts by adults residing in each cell directed towards invading young not born in that cell and the attachment of young to their home cell may be expected to dampen the above process until resettlement in a population maintained at 444 adults would more likely just cover the range of seven to nineteen per cell, harmonising with the range of tolerance in group size

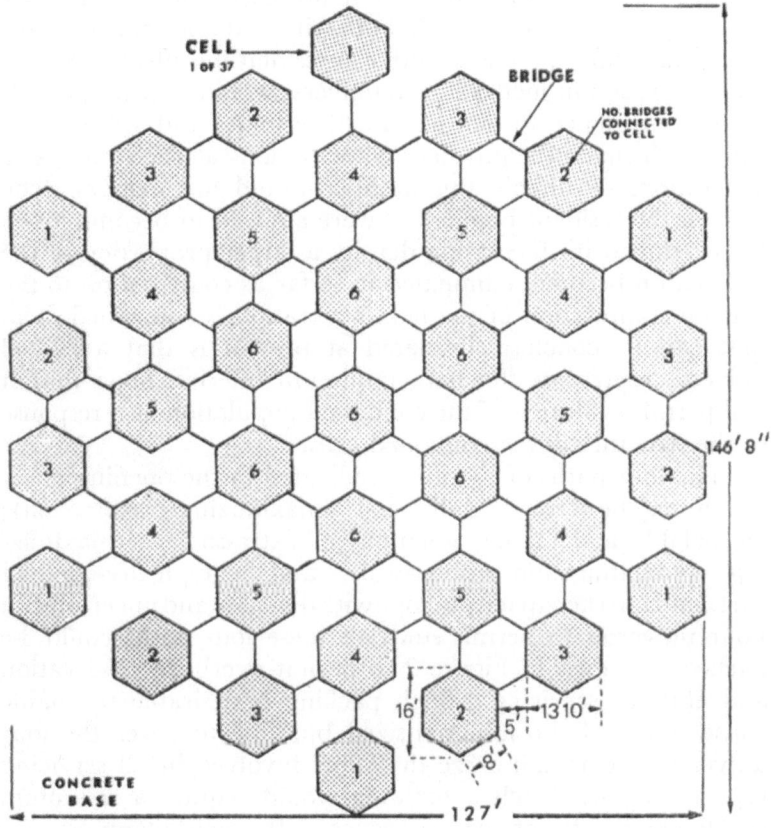

FIGURE 1

37-CELL RAT UNIVERSE

Cell and inter-cell bridge arrangement

compatible with the evolutionary origin of species whose basic group size is twelve individuals.

As a result of the social stratification that develops among the members of any cell, as well as of the equally important social stratification among cells, a very marked differential develops among adults with regard to their genetic contribution to the following generation. For this reason, the effective breeding population is always considerably less than the actual number of adults present. From my experience with social groupings and stratification in both wild and domesticated rats, I can only make a crude intuitive guess that the effective breeding population in such a 37-cell universe would always approximate equivalence to one in which 50 males and 150 females were each making equivalent contributions at each successive generation. Probably this number would not increase even though the resident population were allowed to become much larger, since with increasing density a larger proportion of the population becomes eliminated in so far as contribution to the genetic composition of the next generation is concerned. All that can be conclusively stated at present is that a 37-cell universe represents the bare minimum offering hope that it will permit evolution of the contained population as a response to the structure of the contained cells.

The four paths of evolution indicated in the opening paragraph may be termed ' dull social ' (maximizing gratification); ' a-social ' (maximixing isolation); ' extra-social ' (maximizing aggregating and withdrawal); and ' complexity-coping ' (maximizing the capacity to cope with diversity and uncertainty). Four universes to permit study of these four paths could be packed as shown in Figure 2 to permit overhead observation and efficient servicing. Such packing is desirable to enable construction of the minimal-sized building to cover the four universes and to minimize the effort involved in all servicing type functions. Such a building would require a minimum of 89,000 sq. ft. of ground space. In such planning and simulation study one must keep the practical aspects in view in addition to the theoretical issues from which they stemmed. Along this line, it may be noted that the standing population at any one time for the four universes would average about

FIGURE 2

FOUR-UNIVERSE ARRANGEMENT

⚙ Dull-Social　○ A-Social　● Extra-Social　✫ Complexity-Coping

2500 adults and 2500 immature but weaned rats, and that between 5000 and 10,000 rats would be born each year. Periodic handling of this number of rats would not be a great task were they all domesticated strains confined to small cages; but the increase in reactivity resulting from sojourn in such a rich social and physical environment as would be provided by such universes makes the difficulty of handling and observing even domesticated strains increase several-fold. This difficulty becomes compounded several-fold when it is considered that most individuals will retain a considerable component of genetically determined heightened reactivity as a result of wild type strains having contributed significantly to the originally introduced stock. One might expect to obtain an effective generation in terms of genetic turnover at least every nine months, which gives fifteen generations in ten years. This is probably the minimal number of generations required to produce significant changes from natural selection.* During this time, the one kind of observation of each individual requiring capture and handling would have involved, over this period of time, some 300,000 instances of capturing. These comments give a rough idea of the magnitude of the logistics of such a research effort. I particularly emphasize these practical aspects of such research, because we are approaching a critical era of decision, affecting our role as biologists. The trickle of a few small voices of fifteen or twenty years ago, crying in the wilderness about the threat of the coming increase in the numbers of man, has grown into a torrent during the past five years. The volume of hard data required to generate new insights and validate old hypotheses have not kept pace with the editorial verbiage on populations. I nearly wrote ' garbage ' for ' verbiage '; old ideas and data have been dredged up and re-used so often that

* I use ' natural selection ' to encompass procedures by the investigator which some might consider as ' artificial selection '. For example, in the extra-social universe a criterion would be established that all residents would be removed from every cell maintaining less than a specified number of adults for any three-month period. This is analogous to a situation in nature whereby a group falling below a given size becomes more vulnerable to predation. In essence, criteria will be established for governing the vulnerability of residents within a universe to the ' predatory actions ' of the investigator. These criteria may change over time in order to accelerate the direction of selection imposed by the structure of the environment.

they are becoming stale. The time has come when we, as biologists, have the obligation to begin stating, in fairly precise terms, the magnitude of the effort required to produce those further insights which we have promised as being within the realm of our capability in contributing to the solution of problems relating to population growth, its stabilization and the subsequent continuing evolution of man. Such is the purpose of the present exercise in simulation and suggestion. However, as soon as one does embark on such an effort, he soon realizes the depths of his ignorance, his ineptitude for coping with such complex, long-range studies, and the truly primitive conceptual tools with which his vaunted science has provided him. I hope that you will help me shoulder the anxiety this exploration engenders.

THE FOUNDING POPULATION

Meeting the objective of maximizing the magnitude of the gene pool of the first breeding generation within each universe can be approximately attained by the following strategy. Due to the wide geographic range of the Norway rat and the long isolation of many populations from each other, it should be possible to procure an adequate breeding stock of at least six different wild strains each of which has a good possibility of containing genes which are absent from or are present only in very low frequency in the other strains. Similarly, it is possible to obtain at least eighteen domesticated strains, each differing considerably from the others as to genotype. Each wild type strain would be grouped with three of the domesticated strains to form the basis for a four-way cross providing pregnant females for introduction into the universes. Thus, with regard to the combination of strains involved in the four-way cross, six different types of pregnant females would become available. Twelve females of each of these six types would be introduced into each universe, two females to each cell except for the centre cell of the universe. In this way two females could be permitted to rear litters in each of these cells. To further uniformity of initiation of the four universes, each female would be permitted to rear only six young, while keeping the overall sex ratio as nearly equal as possbile. In this way the initial cohort of similar-aged rats in each universe would consist of 432 individuals. With this background, we may now turn to the character of the internal structuring of cells in each universe capable of providing the setting for the above-named four paths of evolution.

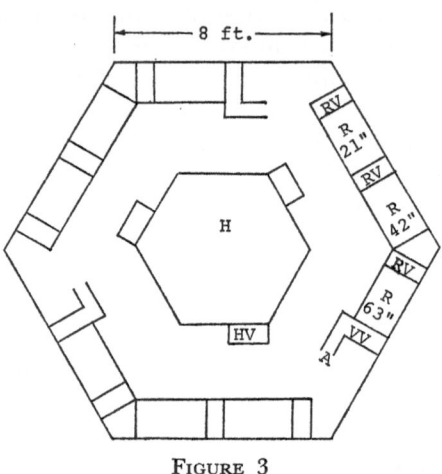

FIGURE 3

DULL-SOCIAL AND COMPLEXITY-COPING UNIVERSE CELL ; HORIZONTAL
COLLAPSED X-SECTION OMITTING MEZZANINE

VV = vertical ramp to mezzanine
A = access tunnel to vertical ramp to mezzanine
H = harbourage unit ; a simulated " burrow " containing 18 nest
     boxes and interconnecting tunnels
HV = vertical ramp to harbourage unit
R = resource shelf for food, water, trait-acquisition situations, etc.
     Shelves at three levels: 21 inches; 42 inches; 63 inches.
RV = vertical ramp to resource shelf

A DULL-SOCIAL PROMOTING UNIVERSE

Given such a species as the Norway rat which has an N_b of approximately twelve adults the objective in constructing the present type of universe is to design the contained cells so that each provides optimum conditions for the preservation of this group size, but in so doing to produce extreme stability of all conditions over time. Figure 3 presents in diagrammatic form a collapsed cross-sectional view of such a cell. In the centre lies a harbourage unit, a simulated burrow system. I have excavated many rat burrows; from a study of these it has been possible to develop plans for a simulated burrow system constructed of metal which contains a system of tunnels and adjoining compartments where the rats may construct nests. The number and arrangement of tunnels and compartments can be made to approximate, in an idealized fashion, that inhabited by a group of rats known to be characterized by the most favourable and stable social structure. The particular details are here unimportant, except to note that it is an elevated hexagonal table-like structure in which the openings into the artificial burrow system are from its top surface. Three vertical ramps connect this surface with the floor. Three sets of three shelves are attached to the walls of the cell opposite the harbourage unit. Each shelf is connected to the floor by a vertical ramp. Each shelf contains identical sources of food, water and nesting material. Adjacent to the end of the higher shelf of each set, a vertical ramp leads from the floor to a mezzanine running around the cell at an eight-foot level above the floor. From this mezzanine, tubes forming bridges connect adjacent cells according to the pattern shown in Figure 1.

Nothing changes in the physical environment from generation to generation. Needed resources may always be found in abundant supply in exactly the same places. Excess population is periodically removed from cells diverging most in numbers, either less than or more than, from ones containing twelve adults. This artificial predation serves both to preserve an uncrowded stable population and to speed up the process of reselection within the diversified gene pool towards a genotype most adapted to a group size of twelve adults in this particular environment. It may be expected that social relations both within and between groups will rapidly achieve a stable pattern, and that this pattern will persist from generation to generation.

In these circumstances the inability to recognize whether or not an associate is in a state of readiness to participate in any given type of interaction will remain as the only significant uncertainty in the environment. Selection may be expected to proceed towards enhancing behaviours, perceptiveness and physical characteristics which will facilitate recognition of the state of readiness. These might take the form of fleshy protuberances capable of changing colour, of ruffs which could be elevated or laid flat, or of stereotyped motor patterns or dances.* Through these signalling devices, each individual entering a state of readiness to begin interacting with another could more readily identify others in a similar state and thus avoid being thrown into the frustrating refractory periods engendered by advances towards individuals already in a refractory state and thus incapable of appropriate response. In the absence of necessity to tolerate the physiological changes accompanying frustration, selection for such tolerance would diminish. This would lead to a lethargic species no longer needing the stimulation produced by frustration. Hedonistic life would flow from one bout of social orgasm to another, rarely punctuated by more than the minimal physical efforts required for obtaining food and water. In all likelihood the amount of time spent in sleep would greatly increase. In this simple, static environment, devoid of uncertainties and challenges, pressures to preserve and accentuate intellectual abilities would largely disappear. Thus the term " dull-social " can best designate the end product of the path of evolution promoted by such an environment.

AN EXTRA-SOCIAL PROMOTING UNIVERSE

The objective here is to produce an environment that will foster maintaining large aggregations. Figure 4 presents in diagrammatic form the arrangement of environmental components required to meet this objective. A cafeteria like arrangement is located in the centre of each cell. Around its hexagonal surface, enclosing the reservoirs for food and water, are located platforms where rats can come to secure these needed resources. At each water platform there are several

* How far the characteristics of the animals within a universe can be pushed in the anticipated direction within the span of fifteen generations will depend in part upon (1) the variability of the initial gene pool, (2) the amount of similar variability introduced over time to compensate for gene drift, and (3) the effectiveness of the investigator as a ' predator ' in accelerating the direction of selection imposed by the physical environment.

levers. To obtain a drop of water a rat must press a lever several times in succession. This constraint requires that each rat spend more time on the water platform, and thus the probability of another rat coming up and pressing at an adjacent lever will increase. Similarly the requirement of having to gnaw at solid food through a narrow slot at each food platform

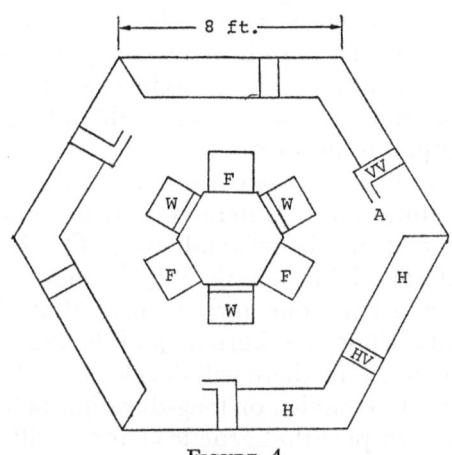

FIGURE 4

EXTRA-SOCIAL UNIVERSE CELL; OVERHEAD ASPECT

VV & A as for Figure 3
   H =Harbourage shelf. Eight nest boxes are suspended under each shelf
   HV =vertical ramps giving access to harbourage shelves at 28-and 56-inch elevations
F & W =platforms at each of which several rats can be side by side while obtaining food or water

will increase the likelihood of rats eating beside each other. This will enhance the tendency for each rat to redefine the eating and drinking situation as one requiring the presence of other rats while engaging in the related behaviours involved in procuring food and water. In their travels between cells, rats will note that in some cells this definition is more likely to be met than in others. The initial reason for this is that rats will tend to congregate in cells in proportion to the number of accesses from adjacent cells. Gradually those cells having more rats, because of this process alone, will begin to acquire additional residents through the activation of the behavioural sink phenomenon. As a result the innermost nineteen cells will average at least two and a half times as many adults as the more peripheral eighteen cells. With the passage of time, the continued operation of the behavioural sink phenomenon will probably even further exaggerate the degree of congregating in the innermost nineteen cells. By acting as a predator, and removing all young born in the less populous cells, a stage can probably be achieved where maximal adaptation is to life in a group containing at least forty-eight adults, four times the original optimum number.

Adaptation to this increased group size will require that the intensity and duration of interacting with others becomes markedly reduced from the original level. On the basis of my observations of rats living out their lives under conditions of excessive congregations, one may suspect that since such a pattern of life also leads to a shortening of the duration of many other types of behaviour, there will develop an adaption to this inability to execute complex or long-duration behaviours. As a result, one may suspect that genetic changes will occur which permit survival despite a much simplified repertoire of behaviours. Having only the ability to express short, simple and probably very stereotyped behaviours suggests the development of a form with little capacity to cope with complexity or uncertainty. Furthermore, the harbourage units designed for such a universe are such as to reduce the chances that adjusting to their structure will facilitate maintaining an ability to cope with complexity. In each cell in this universe, the harbourage units would consist of three sets of three long shelves. Along their length, entrance ways lead directly into single isolated nest compartments. The point to note here is that the complexity of the maze-like pattern of the tunnels connecting nesting compartments in the harbourage unit of the dull-social cells is missing in the cells of the extra-social universe cells. This pervading aura of simplicity, which may be expected to become accentuated in populations inhabiting the extra-social universe, must also lead to an increasing simplification of the entire fabric of social organization.

AN A-SOCIAL PROMOTING UNIVERSE

Following up the insights derived from species which in nature lead an extremely solitary way of life, the objectives of such a universe are to reduce the likelihood of contact between individuals and to reduce the opportunity of contact with diverse and variable stimuli. Figure 5 presents the basic plan for a cell in such a universe. The central small hexagon represents an area from which rats are excluded. This area is

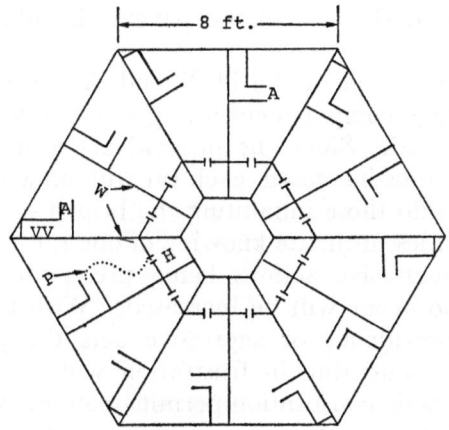

FIGURE 5
A-SOCIAL UNIVERSE CELL; X-SECTION NEAR FLOOR LEVEL

VV & A as for Figure 3.
W = partition between subcells
H = harbourage compartment containing food, water, nest site, nesting material and an activity wheel
P = path between harbourage compartment and access tunnel to ramp to mezzanine

used by the investigator to facilitate provision of food and water into the twelve surrounding small four-sided compartments. Each of these represents a darkened enclosure within which a rat can find all the resources required to meet most basic needs. Here is found a food hopper, a water reservoir, a small nest box and an activity wheel. With the exception of sexual and other social activities there is no need for a rat to leave this small harbourage compartment. At most one could expect only two

G

or three adult rats or a female and her unweaned litter to inhabit such a harbourage compartment. The radial lines from the central small hexagon to the walls of the cell represent high vertical partitions preventing direct access from one of the twelve sub-cells of a cell into an adjoining sub-cell. For a rat to get from the harbourage compartment in one sub-cell to that of another it must enter the bare floor of its own sub-cell and cross to the vertical ramp leading to the mezzanine above the dividing partitions. It must then go around to the top of the vertical ramp of another sub-cell and follow it down, and thence across another open floor space to the other harbourage compartment.

By this arrangement the likelihood of any rat leaving its own harbourage compartment and going to that of another is greatly decreased. Since the interval between contacts with neighbours is thus increased, each rat will know its neighbours less well than do those inhabiting the prior two universes discussed. With less intimate knowing of one rat by another, the chances of aggressive actions being precipitated upon such contacts as do occur will be increased. With this prevailing heightened prevalence of aggressive acts the proportion of interactions culminating in frustration will increase. Those rats whose genetic constitution permits tolerance of the heightened degree of associated stress will leave the most progeny and thus such tolerance to stress will increase through sequential generations.

Certain physical conditions can also facilitate evolution of behaviours further maximizing avoidance. In all cells of all four universes an eighteen inch long, four inch diameter tunnel leads to the base of the vertical ramp which provides access from the floor to the mezzanine. Similarly, in every cell a collection of standardized objects will be provided which include such items as small blocks of wood, sticks and squares of paper simulating leaves. It may be expected that rats in the a-social universe will become particularly prone to plug up the entrances to these tunnels and that such behaviours will be further strengthened through selection. Such actions will help prevent access to any sub-cell by residents of another. The end product of this path of evolution should culminate in forms which are both markedly retiring and markedly aggressive. This process can be speeded up by periodic removal of all residents in the central cells where the largest congregations will develop.

A UNIVERSE PROMOTING COMPLEXITY-COPING

The design of such a universe demands that, as the residents learn ways of coping with one new task, another be added which is more difficult than the preceding. Gross physical configurations can be identical to those of the dull-social universe shown in Figure 3. However, instead of the resources available on each of the nine shelves in each cell being constant, abundant and easily obtainable, each of these shelves will become the site for encountering one or more problems to be solved. Solution of any problem requires acquisition of some new behaviour, which may be called a trait. On every shelf, for every trait, there can be constructed a condition, situation or manipulanda, towards which expression of the appropriate trait reflects an accomplishment in overcoming restraints or exploiting opportunities. Considering that there are 333 shelves in this universe, and that working towards establishing 32 traits within the environment may be a realistic goal, there would be an increase from 333 to 10,656 task situations by the terminal phase of such a study.

No attempt has been made to work out the deatils of such an array of task situations. However, some of their characteristics can be specified in terms of their demands upon the rats; they include:

1. Requiring increasing manual dexterity.
2. Placing constraints on the time and place that a task situation is functional; which leads to:
3. Developing searching behaviour; which under marked constraints demands:
4. Signalling others to participate in the availability of the rare resource.
5. Requiring collaboration.
6. Increasing complexity.

90   INFLUENCES ON BEHAVIOUR

This list could be greatly extended and refined. Whatever types and sequence of task situations might finally be deemed most desirable, the reactions of each individual to each task situation must be at least temporarily recorded, and monitored with the aid of computers which have the capacity to programme the next level of complexity on the basis of the degree to which the members of each group inhabiting a particular cell, have mastered the tasks with which they are faced, or have exploited new opportunities for behaving and adjusting. Following certain recent leads regarding human cultural evolution, the programme for a comparable evolution among rats might be according to the schedule shown in Table 1.

TABLE I

COMPLEXITY-COPING TRAIT ACQUISITION

ACCUMULATED NUMBER OF TRAITS ACQUIRED	NUMBER OF TRAITS ACQUIRED DURING TRAIT-DOUBLING INTERVAL	MONTHS TO DOUBLE	MONTHS/TRAIT ADDED	ACCUMULATED TIME IN MONTHS OF CULTURAL EPOCH
1	1	6	6	6
2	1	9	9	15
4	2	13·5	6·75	21·75
8	4	20·2	5·05	41·95
16	8	30·3	3·8	72·25 (6 yrs.)
32	16	45·5	2·83	117·75 (10 yrs.)

Although one cannot predict exactly how fast it might be possible to push a population along the pathway of adaptations forced by such a schedule, it is certainly quite likely that within the fifteen generations possible in a ten-year study some quite drastic changes in gene frequency would result. This shift could be accelerated by selecting out the residents of cells falling below the average for that generation.

CONCLUDING COMMENTS

The type of simulation effort I have attempted here may be considered as a report on future research. Little precedent exists to guide the format of such a report. And yet, I believe, we must attempt to prepare and publish such reports so that they may be scrutinized by our colleagues over the wide range of disciplines that will become involved in the conduct of any such long-range, complex and costly study. In the past, corrective measures have followed after the fact of both execution and publication of a research project. However, I do not believe that we can justify the expenditure of the large amount of funds required to conduct a programme such as I have outlined without having gone through a succession of such simulations, and at each stage having received detailed comments about the proposals. The present paper represents a summary of a third stage in the development of these ideas over the past several years.* I am particularly indebted to Mr. Gerald G. Wheeler for his contribution towards the development of this stage. At the close of the development of the earlier, second, stage detailed consideration was given to it by a group of invited consultants during a two-day conference, namely:

Kyle R. Barbehenn, Smithsonian Institution (behavioural ecology).

Alexander Kessler, Rockefeller University (physician; population genetics). (Now with WHO, Geneva.)

John A. King, Michigan State University (behavioural ecology).

Paul D. Maclean, NIMH (neurophysiology).

Charles H. Southwick, Johns Hopkins University (population ecology, and social behaviour).

Walter C. Stanley, NIMH (developmental psychology).

Kenneth E. F. Watt, University of California, Davis (systems ecology).

Completion of such a critical examination makes possible a marked clarification of theory, objectives and methods, which through successive similar stages of resolution can lead to a final programme having greater chances of success, if and when it may prove feasible to initiate such studies. My concern here

---

* This third stage is summarized in: John B. Calhoun and Gerald G. Wheeler, *Mammalian Behavioral Evolution*, pp. ii + 20, 2 tables, 24 figures. Feb. 10, 1967. Copies available from the author.

is with the development of a planning strategy for complex, long-range, large-scale biological and social studies which will proceed through an ever-widening circle of exposure to comment and correction. Completion of this process may well justify establishment of journals and monograph series solely devoted to research of the future.

Included in this process is the problem of assuring an adequate scope of coverage of ideas and theories which may be relevant to such complex studies, and of establishing the legitimacy of the planning strategy. As an example of what may be required I will mention only three strategies or involvements which have contributed to the thinking reflected in the present proposals. I was co-organizer of an *ad hoc* committee on Physical and Social Environmental Variables as Determinants of Mental Health. This fifteen member committee includes a wide range of competencies ranging from ecology through sociology, psychology, philosophy and jurisprudence to chemistry, physics and urban planning. We have gathered together for two three-day sessions each year for ten years. At each session, assisted by invited guests, we have concentrated on a wide range of issues affecting the future. During the past two years I have gained many new insights from participating in the Columbia University continuing seminar on Ecology and Cultural Evolution. Then to obtain an in-depth perspective I have abstracted 380 books and articles which focus on theory and problems relating to population dynamics and social behaviour. The 3200 resultant twenty-line excerpts are being subjected to a computer-assisted content-analysis based on an extremely in-depth indexing of contained concepts.

Continuation of such strategy may ultimately justify initiation of studies, such as the one presently proposed to investigate the processes of evolution along four major paths which are open for man to take in his very near future. It is my conviction that only a path such as is here termed ' complexity-coping ' presents an opportunity for man continuing along the evolutionary trajectory so far characterizing his history. Stepping into the bypaths leading to maximizing isolation, aggregation, or gratification is likely to present traps which can only restrict man's opportunity for exploring possible options. I have elsewhere[5, 6], discussed additional aspects of the processes and consequences of involvement in these paths of evolution.

REFERENCES

1. CALHOUN, J. B. 1962. Population density and social pathology. *Scient. Am.* **206,** 139.
2. CALHOUN, J. B. 1962. A " behavioral sink ". In *Roots of Behavior.* Ed. E. L. Bliss. p. 295. New York. Harper Hoeber.
3. CALHOUN, J. B. 1963. *The ecology and sociology of the Norway rat.* U.S. Public Health Service Publication No. 1008.
4. CALHOUN, J. B. 1964. The social use of space. In *Physiological Mammalogy.* Ed. Wm. Mayer and R. Van Gelder. Vol. I, p. 1. New York. Academic Press.
5. CALHOUN, J. B. 1966. Role of space in animal sociology. *J. soc. Issues* **22,** 46.
6. CALHOUN, J. B. 1966. A glance into the garden. In *Three Papers on Human Ecology.* Mills College Assembly Series, 1965–1966. Mills College, Oakland, Cal.
7. CALHOUN, J. B. 1967. Ecological factors in the development of behavioral anomalies. In *Psychopathology—Animal and Human.* Ed. Joseph Zubin and Howard F. Hunt. p. 1. New York. Grune and Stratton.

EFFECTS ON BEHAVIOUR
OF DISRUPTION OF AN AFFECTIONAL
BOND

JOHN BOWLBY

The Tavistock Clinic, London

FAMILY doctors, priests and perceptive laymen have long been aware that there are few blows to the human spirit so great as the loss of someone near and dear. Traditional wisdom knows that we can be crushed by grief and die of a broken heart, and also that a jilted lover is apt to do things that are foolish or dangerous to himself and others. It knows too that neither love nor grief is felt for just *any* other human being, but only for one, or a few, particular and individual human beings. The core of what I term an ' affectional bond ' is the attraction that one *individual* has for another *individual*.

Until recent decades, science has had little to say about these matters. Experimental scientists in the physiological or Hullian learning theory traditions of psychology have never shown interest in affectional bonds, and have sometimes talked and acted as though they do not exist. Psychoanalysts, by contrast, have long recognized the immense importance of affectional bonds in the lives and problems of their patients, but they have been slow to develop an adequate scientific framework within which the formation, maintenance and disruption of such bonds can be understood. The void has been filled by ethologists, starting with Lorenz's classical paper on *The Companion in the Bird's World*,[24] progressing through a multitude of experiments on imprinting[3, 31] to studies of bonding behaviour in sub-human primates,[22, 28] and inspiring psychologists to make similar studies of humans.[2, 29]

PREVALENCE OF BONDING

Before discussing the effects of bond disruption a note about bonding and its prevalence is in place. The work

94

referred to shows that, even if not universal in birds and mammals, strong and persistent bonds between individuals are the rule in very many species. The types of bond that are made differ from one species to another, the commonest being those between one or both parents and their offspring and those between adults of opposite sex. In mammals, including the primates, the first and most persistent bond of all is usually that between mother and young, a bond which often persists into adult life. As a result of all this work, it is now possible to view the strong and persistent affectional bonds made by humans from a comparative standpoint.

Affectional bonding is a result of the social behaviour of each individual of a species differing according to which other individual of his species he is dealing with; which entails of course an ability to recognize individuals. Whilst each member of a bonded pair tends both to remain in proximity to the other and to elicit proximity-keeping behaviour in the other, individuals who are not bonded show no such tendencies; indeed, when two individuals are not bonded, one often strongly resists any approach the other may attempt. Examples are the attitudes of a parent towards the approach of young not its own, and of a male towards the approach of another male.

The essential feature of affectional bonding is that the two partners tend to remain in proximity to one another. Should they for any reason be apart, each will sooner or later seek out the other and so renew proximity. Any attempt by a third party to separate a bonded pair is strenuously resisted : not infrequently the stronger of the partners attacks the intruder whilst the weaker flees, or perhaps clings to the stronger partner. Obvious examples are situations in which an intruder is attempting to remove young from a mother, e.g. calf from cow, or to detach the female from a bonded heterosexual pair, e.g. goose from gander.

A little paradoxically, behaviour of an aggressive sort plays a key role in maintaining affectional bonds. It takes two distinct forms : first, attacks on and frightening away of intruders and, secondly, the punishment of an errant partner, be it wife, husband or child. There is evidence that much aggressive behaviour of a puzzling and pathological kind originates in one or other of these ways.[5]

Affectional bonds and subjective states of strong emotion tend to go together, as every novelist and playwright knows. Thus, many of the most intense of all human emotions arise during the formation, the maintenance, the disruption and the renewal of affectional bonds—which, for that reason, are sometimes called emotional bonds. In terms of subjective experience, the formation of a bond is described as falling in love, maintaining a bond as loving someone, and losing a partner as grieving over someone. Similarly, threat of loss arouses anxiety and actual loss causes sorrow ; while both situations are likely to arouse anger. Finally, the unchallenged maintenance of a bond is experienced as a source of security, and the renewal of a bond as a source of joy. Thus, anyone concerned with the psychology and psychopathology of emotion, whether in animals or man, is soon confronted by problems of affectional bonding : what causes bonds to develop and what they are there for, and especially the conditions that affect the form their development takes.

In so far as psychologists and psychoanalysts have attempted to account for the existence of affectional bonds, the motives of food and sex have almost always been invoked. Thus in attempting to explain why a child becomes attached to his mother, both learning theorists [12, 30] and psychoanalysts [14] have independently assumed that it is because mother *feeds* child. In attempting to understand why adults become attached to one another, sex has commonly been seen as the obvious and sufficient explanation. Yet, once the evidence is scrutinized, these explanations are found wanting. There is now abundant evidence that, not only in birds but in mammals also, young become attached to mother-objects despite not being fed from that source,[9, 19] and that by no means all affectional bonding between adults is accompanied by sexual relations ; whereas, conversely, sexual relations often occur independently of any persisting affectional bonds.

What is now known of the ontogeny of affectional bonds suggests that they develop because the young creature is born with a strong bias to approach certain classes of stimuli, notably the familiar, and to avoid other classes, notably the strange. As regards function, observation of animals in the wild strongly suggests that the biological function of much, if not all, bonding is protection from predators—a function fully as important for the survival of a population as nutrition or reproduction, but one which has habitually been overlooked by workers confined in laboratories and concerned only with man living in economically developed societies.

Whether these hypotheses are supported by further work or not, an individual's capacity to make affectional bonds of a type appropriate to each phase of his species' life cycle and to his or her own sex is plainly a capacity as typical of individuals of mammalian species as are their capacities, for example, to see, to hear, to eat and to digest. And in all likelihood a capacity for bonding has as high a survival value to a species as has any of these other long-studied capacities. It is proving productive to view many of the psychoneurotic and personality disturbances of humans as being a reflection of a disturbed capacity for making affectional bonds, due either to faulty development during childhood or to subsequent derangement.

DISRUPTED BONDS AND PSYCHIATRIC ILLNESS

Those who suffer from psychiatric disturbances, whether psychoneurotic, sociopathic or psychotic, always show impairment of the capacity for affectional bonding, an impairment that is often both severe and long lasting. Although in some cases this impairment is clearly secondary to other changes, in many it is probably primary and derives from faulty development having occurred during a childhood spent in an atypical family environment. Whilst disruption of the bonds that tie a child to his parents is not the only form, adverse in this respect, that the environment can take, it is the form most reliably recorded and the effects of which we know most about.*

In considering the possible causes of psychiatric disturbance in childhood, child psychiatrists were early aware that antecedent conditions of significantly high incidence are either an

* There are also valuable studies of the reaction of adults to bereavement and of the relationship of bereavement reactions to mental illness.[27] In a short paper it has not been possible to include discussion of these findings.

98   INFLUENCES ON BEHAVIOUR

absence of opportunity to make affectional bonds or else long
and perhaps repeated disruptions of bonds once made.[1, 4]
Though the view that such conditions are not only associated
with subsequent disturbance but are causal of it is widely held,
that conclusion nevertheless remains debatable.

Studies of the incidence of childhood loss in different
samples of psychiatric populations have multiplied in recent
years. Due to the samples and the comparison groups being
so differently constituted, to criteria of loss being differently
defined, and to a host of demographical and statistical hazards,
their interpretation is not easy. Certain findings, however,
have been so consistently reported by independent workers,
including reports of a number of recent and well-controlled
studies, that we can be reasonably confident of them. Two
psychiatric syndromes and two sorts of associated symptom are
consistently found to be preceded by a high incidence of disrupted affectional bonds during childhood. The syndromes
are psychopathic (or sociopathic) personality and depression ;
the symptoms persistent delinquency and suicide.

The *psychopath* (or *sociopath*) is a person who, whilst not
being psychotic or mentally subnormal, persistently engages
in : i. acts against society, e.g. crime ; ii. acts against the
family, e.g. neglect, cruelty, sexual promiscuity or perversion ;
iii. acts against himself, e.g. addiction, suicide or attempted
suicide, repeatedly abandoning his job.

In such people the capacity to make and maintain affectional bonds is always disordered and not infrequently conspicuous by its absence.

More often than not the childhoods of such individuals are
found to have been grossly disturbed by the death, divorce or
separation of the parents, or by other events resulting in disruption of bonds, with an incidence of such disturbance far
higher than is met with in any other comparable group,
whether drawn from the general population or from psychiatric
casualties of other sorts. For example, in a study of well over
a thousand consecutive psychiatric out-patients under the age
of sixty, Earle and Earle [13] diagnosed sixty-six as sociopaths
and 1357 as suffering from some other disorder. Taking as
their criterion an absence of the mother for six months or more

before the sixth birthday, Earle and Earle found an incidence of 41 per cent for the sociopaths and 5 per cent for the remainder.

When the criterion is made broader the incidence rises. Thus Craft *et al.*[10] took as their criterion an absence of either mother or father (or both) before the child's tenth birthday. Of seventy-six male inmates of the special hospitals for aggressive psychopaths, no less than 65 per cent had had such an experience. In a study of several comparison groups Craft shows how the incidence of this type of childhood experience rises with the degree of antisocial behaviour shown by a group's members.

Others who have reported similar sorts of statistically significant findings for groups of psychopaths and persistent delinquents are Naess,[26] Greer,[15] Brown and Epps ;[7] and for alcoholics and addicts, Dennehy.[11]

In psychopaths the incidence of illegitimacy and a shunting of the child from one ' home ' to another is high. It is no accident that Brady of the ' Moors ' murders was such a one.

Another psychiatric group which shows a much raised incidence of childhood loss is that of *suicidal patients*, both those who attempt it and those who succeed.* The losses are especially likely to have occurred during the first five years of life and to have been caused not only by the death of a parent, but also by other long-lasting causes, notably illegitimacy and divorce. In these respects suicidal patients tend to resemble sociopaths and, as will be seen later, to differ from depressives.

Of the many studies reporting a very high incidence of childhood loss among attempted suicides, e.g. Bruhn,[8] Greer,[16] Kessel,[23] a recent study by Greer, Gunn and Koller [18] is among the best-controlled. A series of 156 attempted suicides were compared with similar-sized samples of non-suicidal psychiatric patients and of surgical and obstetric patients without a psychiatric history ; both comparison groups were matched with the attempted suicides in respect of age, sex, class and

* Although any group of suicides and attempted suicides will contain some sociopaths and some depressives, a majority are likely to be diagnosed as suffering from neurosis or personality disorder [18] and so constitute a fairly distinct psychiatric group.

other relevant variables.   Taking as his criterion of loss the continuous absence of one or both parents for at least twelve months, Greer finds that such events have occurred before the fifth birthday three times as often in the group of attempted suicides as in either of the comparison groups—an incidence of 26 per cent against 9 per cent for each of the others (Table 1).

TABLE I

INCIDENCE OF LOSS OR CONTINUOUS ABSENCE OF
ONE OR BOTH NATURAL PARENTS FOR AT LEAST 12 MONTHS
BEFORE THE 15TH BIRTHDAY

	NON-PSYCHIATRIC PATIENTS	NON-SUICIDAL PSYCHIATRIC PATIENTS	ATTEMPTED SUICIDE
	%	%	%
Age at loss			
0–4 years	9	9	26
5–9 years	12	10	11
10–14 years	7	7	11
Doubtful	0	2	1
0–14	28	28	49
N	156	156	156

Furthermore, the losses in the attempted suicide group tended more often to have been of both parents and to have been permanent, whereas in the other groups they more often concerned only one parent and were temporary, having been due to such exigencies as illness or work.

In a further study of the same group of attempted suicides [17] it was found that those who had suffered parental loss before their fifteenth birthday differed significantly in certain respects from those who had not.   One such difference, in keeping with other findings, is that those who had suffered childhood loss were more likely to be diagnosed as sociopaths than were those who had not suffered a childhood loss (18 per cent against 4 per cent).

Another condition which is associated with a significantly raised incidence of childhood loss is *depression*.   The type of loss experienced, however, tends to be of a kind different from the overall family disruption typical of the childhoods of psychopaths and attempted suicides. First, in the childhoods of depressives loss is likely to have been due to the death of a parent rather than to illegitimacy, divorce or separation. Secondly, although in depressives the incidence of bereavement tends to be raised during each quinquennium of childhood, losses tend to have occurred especially frequently when the patient was aged between ten and fifteen years. Thirdly, the incidence of loss is apt to be most raised for the parent of the opposite sex—mother in the case of boys and father in the case of girls. Findings of this sort have been reported by Brown,[6] Munro,[25] Dennehy,[11] and Hill and Price.[21] These studies indicate that loss of a parent by death between the patient's tenth and fifteenth birthdays tends to have occurred about twice as frequently in a group of depressives as it has in the population as a whole.

Thus, it now seems reasonably certain that in several groups of psychiatric patients the incidence of disruption of affectional bonds during childhood is significantly raised. Whilst these later studies confirm the earlier findings regarding the raised incidence of loss of mother during early childhood, they also extend them. For several sorts of condition raised incidences of disrupted bonds are now seen to include bonds to fathers as well as to mothers, and to be found during the years from five to fourteen as well as during the first five. Furthermore, in the more extreme conditions, sociopathy and suicidal tendencies, not only is an initial loss likely to have occurred early in life but it is likely also to have been both a permanent loss and to have been followed by the child experiencing repeated shifts of parent figures.

Nevertheless, to demonstrate a raised incidence of some factor is only one thing : to demonstrate that it plays a causal role is quite another. While most of those reporting the findings referred to believe that the raised incidence of childhood loss bears a causal relation to the subsequent psychiatric disturbance, and there is a wealth of clinical records pointing in that direction (for references see Bowlby[5]), alternative explanations still remain possible. As an example, the raised incidence of maternal and paternal death in psychiatric patients might be a result of the parents of patients being older than average at the time of the patient's birth. Were this so, not only would early death of a parent be more likely but there might also be greater liability for the offspring to be born with an adverse genetic loading. Thus, what appears to be an environmental determinant might turn out to be a genetic one after all.

To test that possibility is not easy. For it to be supported requires : first, that the mean ages of the mothers and/or fathers of psychiatric patients be found in fact to be higher than the means for the population as a whole ; and, secondly, that any raised parental age that may be found be shown to have an adverse effect on the genetic endowment of the off-spring such that the likelihood of psychiatric disability is raised. The first requirement may well be met : recent evidence [11] suggests that mean ages of the parents of psychiatric patients may be above those of the population from which they come. The second requirement, however, is more difficult to get evidence about. Plainly it may be some time before the issue is settled.

Meanwhile, those who believe that the relationship between disruption of affectional bonds during childhood and disturbance of the capacity to maintain affectional bonds typical of personality disorders of later life is a causal one point to other evidence in support of their hypothesis. It concerns the way young humans and sub-human primates behave when an affectional bond is broken by separation or death.

SHORT-TERM EFFECTS OF DISRUPTED BONDS

When a young child finds himself with strangers and without his familiar parent figures, not only is he intensely distressed at the time but his subsequent relationship with his parents is impaired, at least temporarily. The behaviour seen in two-year-olds during and after a short stay in a residential nursery is the subject of a systematic descriptive and statistical study undertaken at the Tavistock by Heinicke and Westheimer.[20] The particular part of their report to which I draw attention is that in which they compare the behaviour towards mother of ten children who had been away in the nursery and are now returned home, with that of a comparison group of ten young children who had remained at home throughout.

In the separated children two forms of disturbance of affectional behaviour were seen, neither of which was observed in the comparison group of non-separated children. One form is that of emotional detachment ; the other its apparent opposite, namely an unrelenting demand to be close to mother.

i. On first meeting his mother after he has been away from home with strangers for two or three weeks a two-year-old child typically remains distant and detached. Whereas during his early days away a child commonly cries pathetically for his mother, when at last she returns he seems not to recognize her or avoids her. Instead of rushing to her and clinging as he probably would if he had been lost in a shop for half an hour, he often looks right through her and refuses her hand. All the proximity-seeking behaviour typical of an affectional bond is missing, usually to the mother's intense distress ; and it remains missing—sometimes only for minutes or hours but sometimes for days. Resumption of attachment may be sudden, but is often slow and piecemeal. The length of time detachment persists is positively correlated with the length of the separation (Table II).

TABLE II

NUMBER OF SEPARATED AND NON-SEPARATED CHILDREN WHO SHOWED DETACHMENT DURING FIRST 3 DAYS AFTER REUNION (OR DURING EQUIVALENT PERIOD) [20]

	SEPARATED	NON-SEPARATED
No detachment	—	10
Detachment for one day only	1	—
Detachment alternating with clinging	4	—
Detachment persistent for 3 days	5	—
	10	10

Degree of detachment correlated with length of separation : $r = 0.82$; $P = 0.01$.

H

ii. When—as is usual—attachment behaviour is resumed, a child is commonly much more clinging than he was before the separation. He dislikes his mother leaving him and tends either to cry or to follow her round the house. How this phase evolves turns largely on how his mother responds. Not infrequently conflict ensues, a child demanding his mother's constant company and she refusing it. Such refusal readily evokes hostile and negative behaviour from the child, which is apt to try his mother's patience still further. Of the ten separated children observed by Heinicke and Westheimer six showed strong and persistent hostile behaviour to mother and negativism after their return home : no such behaviour was seen in the non-separated children (Table III).

TABLE III

Number of Separated and Non-separated Children who showed
strong and persistent Hostility to Mother after Reunion
(or during Equivalent Period) [20]

	Separated	Non-separated
Little or no hostile behaviour or negativism to mother	4	10
Strong and persistent hostile behaviour and negativism to mother	6	0
	10	10

$P = 0.01$

Clearly it is still a far cry between showing that a child's bonds to his mother, and often to his father also, are thrown into disequilibrium by a brief separation, and demonstrating unequivocally that long or repeated separations are causally related to subsequent personality disorders. Yet the detached behaviour so typical of young children after a separation bears more than a passing resemblance to the detached behaviour of some psychopaths, whilst it would be difficult to distinguish the aggressively demanding behaviour of many a young child recently reunited with his mother from the aggressively demanding behaviour of many hysterical personalities. To postulate that in each type of case the disturbed behaviour of the adult represents a persistence over the years of deviant patterns of bonding behaviour that have become established as a result of bond disruptions occurring during childhood proves useful. On the one hand it helps to organize data and to orient further research ; on the other it provides guide-lines for the day-to-day management of these kinds of people.

To advance our knowledge in this field it would obviously be invaluable to conduct a long series of experiments to investigate the short- and long-term effects on behaviour of disrupting an affectional bond, taking into account the subject's age, the nature of the bond, the length and frequency of disruptions, and many other variables besides. Equally obvious, however, is that any such experiments on human subjects are ruled out on ethical grounds. For these reasons it is much to be welcomed that comparable experiments using sub-human primates are now being undertaken. Preliminary findings suggest that the effects on six-month-old rhesus infants of a temporary loss of mother (six days) are, both during and after the separation, not unlike those on two-year-old human children,[32] e.g. distress and a lowered level of activity during the separation, and an exceptionally strong tendency to cling to mother after it is over. The monkey-mother's reactions to this, moreover, are not unlike the human mother's. To date, however, there is no record of a monkey baby showing detachment, and this may represent a species difference.

Both in human and in monkey children there are very wide individual variations in the reaction to disruption of a bond. Some of this variation is probably due to the effects on an infant of events occurring during pregnancy and birth. Thus Ucko[33] found that boys who at birth had been recorded as having been in an asphyxiated state are very much more sensitive to environmental change, including separation from mother, than boys who at birth had not been asphyxiated (Table 4). Some other part of this variance, on the other hand, may well be genetically determined. Indeed, it is a reasonable hypothesis that a principal way in which genetic

TABLE IV

DISTRESS AT TEMPORARY SEPARATION FROM MOTHER, FATHER OR SIBLING IN YOUNG BOYS ANOXIC AT BIRTH AND NON-ANOXIC AT BIRTH [33]

	2ND YEAR		3RD YEAR	
	DISTRESSED	NOT	DISTRESSED	NOT
Anoxic	8	2	9	2
Non-anoxic	2	12	4	7
Significance	$P = ·01$		$P = ·1$	

Total samples comprise 29 pairs of boys matched for class, birth order and maternal age

factors act to influence the development of mental health and ill-health is by their effect on bonding behaviour : in what degree and form, and in what circumstances, can an individual make and maintain affectional bonds, and how does he respond to disruption of bonds. By undertaking studies of this sort it may in future be possible to bring together environmental and genetic studies of behavioural disorder.

REFERENCES

1. AINSWORTH, M. D. 1962. *The Effects of Maternal Deprivation: A review of findings and controversy in the context of research strategy.* Geneva. WHO Public Health Papers No. 14.
2. AINSWORTH, M. S. 1967. *Infancy in Uganda: Infant Care and the Growth of Attachment.* Baltimore, Md. Johns Hopkins University Press.
3. BATESON, P. P. G. 1966. The characteristics and context of imprinting. *Biol. Rev.* **41,** 177.
4. BOWLBY, J. 1951. *Maternal Care and Mental Health.* Geneva. WHO. Monograph Series No. 2.
5. BOWLBY, J. 1963. Pathological mourning and childhood mourning. *J. Am. psychoanal. Ass.* **11,** 500.
6. BROWN, F. 1961. Depression and childhood bereavement. *J. ment. Sci.* **107,** 754.

7. BROWN, F. and EPPS, P. 1966. Childhood bereavement and subsequent crime. *Br. J. Psychiat.* **112**, 1043.
8. BRUHN, J. G. 1962. Broken homes among attempted suicides and psychiatric out-patients : a comparative study. *J. ment. Sci.* **108**, 772.
9. CAIRNS, R. B. 1966. Attachment behaviour of mammals. *Psychol. Rev.* **73**, 409.
10. CRAFT, M., STEPHENSON, G. and GRANGER, C. 1964. The relationship between severity of personality disorder and certain adverse childhood influences. *Br. J. Psychiat.* **110**, 392.
11. DENNEHY, C. M. 1966. Childhood bereavement and psychiatric illness. *Br. J. Psychiat.* **112**, 1049.
12. DOLLARD, J. and MILLER, N. E. 1950. *Personality and Psychotherapy.* New York. McGraw-Hill.
13. EARLE, A. M. and EARLE, B. V. 1961. Early maternal deprivation and later psychiatric illness. *Am. J. Orthopsychiat.* **31**, 181.
14. FREUD, S. 1938. *An Outline of Psychoanalysis.* (In Standard Edition Vol. 23.) London. Hogarth Press, 1964.
15. GREER, S. 1964. Study of parental loss in neurotics and psychopaths. *Archs. gen. Psychiat.* **11**, 177.
16. GREER, S. 1964. The relationship between parental loss and attempted suicide : a control study. *Br. J. Psychiat.* **110**, 698.
17. GREER, S. and GUNN, J. C. 1966. Attempted suicides from intact and broken parental homes. *Br. med. J.* **2**, 1355.
18. GREER, S., GUNN, J. C. and KOLLER, K. M. 1966. Aetiological factors in attempted suicide. *Br. med. J.* **2**, 1352.
19. HARLOW, H. F. and HARLOW, M. K. 1965. The Affectional systems. In *Behaviour of Non-human Primates.* Vol. 2. Eds. Schrier, Harlow and Stollnitz. New York and London. Academic Press.
20. HEINICKE, C. M. and WESTHEIMER, I. J. 1966. *Brief Separations.* New York. International Universities Press ; London. Longmans.
21. HILL, O. W. and PRICE, J. S. 1967. Childhood bereavement and adult depression. *Br. J. Psychiat.* **113**, 743.
22. HINDE, R. A. and SPENCER-BOOTH, Y. 1967. The behaviour of socially living rhesus monkeys in their first two and a half years. *Anim. Behav.* **15**, 169.
23. KESSEL, N. 1965. Self-poisoning. *Br. med. J.* 1965. **2**, 1265; 1336.
24. LORENZ, K. Z. 1935. Der Kumpan in der Umwelt des Vogels. *J. Orn., Lpz.* **83** (English translation : In *Instinctive Behaviour*, Ed. Schiller, New York. International Universities Press, 1957).
25. MUNRO, A. 1966. Parental deprivation in depressive patients. *Br. J. Psychiat.* **112**, 443.
26. NAESS, S. 1962. Mother-child separation and delinquency : further evidence. *Br. J. Crim.* **2**, 361.
27. PARKES, C. M. 1965. Bereavement and mental illness. *Br. J. med. Psychol.* **38**, 1.

28. SADE, D. S. 1965. Some aspects of parent-offspring and sibling rela-
    tions in a group of rhesus monkeys, with a discussion of grooming.
    Am. J. phys. Anthrop. **23,** 1.
29. SCHAFFER, H. R. and EMERSON, P. 1964. The development of social
    attachments in infancy. *Monogr. Soc. Res. Child Dev.* **29,** 1.
30. SEARS, R. R., MACCOBY, E. E., and LEVIN, H. 1957. *Patterns of Child
    Rearing.* Evanston, Ill. Row, Peterson.
31. SLUCKIN, W. 1964. *Imprinting and Early Learning.* London. Methuen.
32. SPENCER-BOOTH, Y. and HINDE, R. A. 1966. The effects of separating
    rhesus monkey infants from their mothers for six days. *J. Child
    Psychol. Psychiat.* **7,** 179.
33. UCKO, L. E. 1965. A comparative study of asphyxiated and non-
    asphyxiated boys from birth to five years. *Devel. Med. Child Neurol.*
    7, 643.

INTERRELATIONS

Chairman: PROFESSOR J. M. THODAY

INTRODUCTION

J. M. THODAY

Department of Genetics, University of Cambridge

'INTERRELATIONS', as the title of this session, indicates that the session will involve some consideration of the interrelations of environmental and genetic influences on behaviour. In other words, the title should draw attention to what we call ' genotype-environment interaction ', a technical term with a precise meaning which I spent some time spelling out in a previous symposium of the Eugenics Society.* In fact, our interest in behaviour genetics, and, though perhaps to a lesser extent, our interest in environmental influences on behaviour, must be very largely an interest in genotype-environment interaction.

Behaviour involves reaction to environment. Behaviour genetics is therefore concerned with heritable variation in reaction to environmental variables, which is genotype-environment interaction, the term we use to describe the situation when different genotypes react differently to a particular environmental difference. For instance, red-green colour blindness is a heritable condition which we detect primarily because the individuals concerned react differently from others to the environmental difference between red and green light.

This is a simple example. A more complex one involves capacity for spatial visualization. It has been a regular experience in my teaching career to find a small proportion of students who seem totally incapable of visualizing the three-dimensional shape of a cell from observation of two-dimensional sections taken at right angles to one another. One can get them to cut sections from macroscopic plasticine models but they still seem unable to learn to see the relationship between

* Thoday, J. M. 1965. Geneticism and Environmentalism. In *Biological Aspects of Social Problems*. Ed. J. E. Meade and A. S. Parkes. Edinburgh. Oliver and Boyd.

the sections and the whole. Such people are most unlikely to have success in careers that require this ability. Now there is good evidence that performances in various tests designed to discriminate between people with respect to this sort of spatial visualization have heritabilities as high as 89 per cent. Here we have a situation where career success must depend very much on genotype-environment interaction. It is little use expecting those with low spatial visualization to make a success in the environment of a laboratory of comparative morphology, though in another environment they may not suffer at all, and might even profit from this, largely genetic, handicap. In other words, career success will involve genotype-environment interaction.

In considering interrelations between genetic and environmental influences on behaviour, indeed in considering the whole subject of this Symposium we must therefore always bear in mind that if we find an environmental variable that affects a particular behaviour pattern, we have in no way shown that genetics is irrelevant. Likewise, if we find relevant genetic factors, we have in no way shown that environmental variables are irrelevant. Even if all delinquents could be shown to have suffered from maternal deprivation, it would still be possible that certain combinations of genes were prerequisite for delinquency, unless it could be shown that *every* individual who suffers from maternal deprivation becomes delinquent.

These interrelations become more important and more complex when we remember that genotype can directly determine environment. Of course environment determines genotype indirectly, through natural selection ; but genetic differences can determine environmental differences quite directly. A tall pea meets a different environment from a dwarf one. Red-green colour blind people live in a different environment from colour-perceptive people. Professor Kalmus demonstrated this in a film which he showed at the meeting and we are glad to be able to include his description of the film in this volume.

On the whole people hate having this kind of fact thrust home. I have seen small boys reduced to blows in an argument about whether PTC has a nasty taste or not, before they would accept the fact that for some there is no difference in taste between PTC solution and water. Yet we must face the implications of this kind of fact. People differ, and it is part of our humanity that each of us must live in a different world. Study of the interrelation of genetic and environmental influences on behaviour should help us to understand one another better. Without such understanding we shall not easily live in harmony ; and we shall not reach such understanding if we think that genetic and environmental causes are alternatives. Let us always remember they are complementary.

CHARACTERISTICS OF DELINQUENT BOYS AND THEIR HOMES

J. W. B. Douglas and J. M. Ross

Medical Research Council Unit, London School of Economics

FORMIDABLE problems are encountered when attempts are made to link early experience with later delinquent behaviour. Popular views of the causes of delinquency gleaned from newspaper reports of current theories are likely to have a selective influence on recall. It is widely known, for example, that early separation of a child from his parents is a suspected cause of delinquent behaviour, and so the mothers of children who are in trouble are likely to search their memories for instances of separation in a way that the mothers of other children are not. A further complication is that some suspected antecedents of delinquent behaviour, such as the divorce or separation of the parents, are often followed by a profound deterioration in living conditions which may affect the child by their sharp contrast with the life that has gone before and by exposing him to a relatively sudden change in community values.

The problem of selective recall can be eliminated if delinquent behaviour is studied in a group kept under observation for a long period of time—preferably from infancy to adolescence. The members of such a group have, of course, to be selected before they are of an age to be delinquent. The comparative rarity of delinquent acts, even among boys, and the great variety of offences they commit—varying from such trivial ones as riding bicycles without lights to inflicting serious bodily harm—sets further conditions. Either the definition of delinquency has to be extended to include anti-social behaviour which is not known to the police, or a large sample has to be used. By re-defining delinquency we make assumptions, which may not be valid, as to the relationship between officially recorded delinquency and anti-social behaviour. On the other hand, a considerable increase in sample size is likely to lead to problems of data collection unless large sums of money are available and large numbers of highly skilled research workers can be recruited.

What can be done with relatively modest funds and a relatively large sample is described below. This is, however, the description of work that is still in progress and the results, particularly those relating delinquency with family insecurity, should be regarded as preliminary in the sense that they cover only part of the data available at a rather superficial level.

The sample of young people to be described is scattered throughout Great Britain and its members were all born in the first week of March 1946. From the total births during that week in Great Britain all children whose fathers were in non-manual work or in agricultural work were selected, and one quarter of the remainder. This gave a sample of 5362 boys and girls, approximately half of whom were middle class and half manual working class. This is a convenient sample for many types of study but has the disadvantage in the present context that it cuts down by one-fifth the numbers who are delinquent. The original social structure of the week's births can easily be regained by a simple weighting procedure, and this is used when population estimates are required. By comparison with known national statistics, the population estimates obtained in this way appear to be highly reliable and there is much evidence reported elsewhere that this sample of young people is still, after twenty-one years, highly representative of all those born in 1946.

This paper is concerned with delinquency only among boys, since delinquent acts were so infrequent among the girls that the numbers were insufficient for a detailed study. A delinquent in this discussion is a boy or girl who between the ages of eight and seventeen years has been sentenced by the courts for an offence or has been cautioned by the police. Of the boys in this sample, 288 were delinquent by this definition, and when adjustments are made for the original sampling procedure a delinquency rate of 15 per cent is obtained for this age period. Only 35 girls were delinquent.

The delinquent boys have been divided into three groups,

according to the severity of their offence:

1. *The trivial offenders*—boys who were either cautioned by the police or sentenced for a non-indictable offence ($N = 96$).
2. *The serious offenders*—boys who had committed one, but not more than one, indictable offence ($N = 125$).
3. *The repeaters*—boys who had been sentenced by the courts on more than one occasion for an indictable offence ($N = 67$).

Trivial offenders were reported particularly frequently during the last years at school, that is to say between ages fourteen and seventeen.

It is certain that much delinquent behaviour remained undiscovered, and indeed both the mothers and teachers have informed us of thefts and other anti-social acts that were not known to the police. We have, however, not included these in our definition of delinquency because it is certain that the reporting is incomplete and indeed that many delinquent acts are known only to the child concerned.

The risks of being cautioned by the police or sentenced by the courts are considerably higher in manual working class than in middle class families. Within the manual working class, however, there is no clear relationship between the level of skill of the father's work and risks of delinquency to his children. In this study (see Table 1 *a*) the highest risk of repeated offences was among those boys whose fathers were in semi-skilled rather than in unskilled employment, while among the agricultural workers the risk of delinquency was low.

If the manual workers are divided up, not by the level of employment of the father, but the education and social origins of both parents (see appendix for a definition of the four social classes) a more meaningful picture appears. The risks of delinquency were low when either the father or the mother came from a middle-class background or when either parent had stayed on at school to fifteen or later. In contrast the risks were high when both the father and the mother had manual working class origins and when both had only the minimum of formal education. This is the group labelled lower manual working class in Table I *b*, where it will be seen that these families have nearly twice the proportion of delinquents found in any other

TABLE I

THE PERCENTAGE DELINQUENCY RATES OF THE BOYS GROUPED BY :
a, FATHER'S OCCUPATION ; *b*, SOCIAL CLASS

TYPE OF DELINQUENT

a. Father's Occupation 1961	ALL	TRIVIAL	SERIOUS	REPEATERS	NUMBER
Professional	5·0	3·3	1·3	0·4	241
Salaried	4·5	2·5	1·8	0·2	401
Black-Coated	10·3	5·2	4·2	0·9	331
Foremen and Skilled	15·1	4·3	7·0	3·8	716
Semi-Skilled	20·1	4·6	7·3	8·2	194
Agricultural	14·6	6·0	5·3	3·3	151
Unskilled	20·6	4·2	9·7	6·7	165
Self-Employed	9·0	2·5	5·5	1·0	199
Not Classified	†	Nil	†	†	4
TOTAL	12·0	4·0	5·2	2·8	2,402
b. Social Class					
Middle { Upper	2·7	1·9	0·8	nil	260
Middle { Lower	8·3	4·1	3·5	0·7	738
Manual { Upper	9·7	3·6	4·3	1·8	392
Manual { Lower	18·7	4·9	8·2	5·6	942
Not Classified	8·6	†	†	†	70
TOTAL	12·0	4·0	5·2	2·8	2,402

† = Insufficient numbers

group and more than three times the proportion of repeaters. Even within the lower manual working class group, the risks of delinquency are greatly reduced if one or other of the parents, after having left school at the minimum statutory leaving age, attempted to get further education at night school, through a correspondence college, etc. In families selected in this way, the risks of delinquency (11 per cent) are approximately half those found in the rest of the lower manual working class (20 per cent), and this is so whether it is the father or the mother who has tried to get further education and whether or not their attempts have resulted in some formal qualification.

Thus, any effort on the part of the parents to gain further education is associated with relatively low risks of delinquency in the next generation. This association is slight for trivial offences, but considerable for serious offences, and in particular for repeated offences.

In the early years (see Table II) the parents of the delinquent boys were consistently picked out as giving poor care, failing to use the available medical services and taking little interest in their sons' educational progress.

TABLE II

SOME CHARACTERISTICS OF THE FAMILIES OF THE DELINQUENTS
(Percentage Table)

TYPE OF DELINQUENT

	TRIVIAL	SERIOUS	REPEATERS	EXPECTED*
Poor maternal care	41†	59	78	41
No post-natal services used	15	23	32	17
Low parental interest in child's school work	54	65	70	45
Large family (4 or more children)	42	44	71	36
Overcrowded house (at 15 yrs.)	38	47	73	37
Shared bed (at 11 yrs.)	59	67	79	59

* These percentages were obtained by applying the social class composition of the delinquents to the whole sample of boys.
† i.e. among the trivial offenders 41 per cent were given poor maternal care.

Health visitors called at the homes of the survey members two months after their birth and on two further occasions during the pre-school years. They were asked to assess during these visits what level of care the mothers were giving to their children and to their homes. When later delinquent behaviour is related to these early assessments the trivial offenders are found to receive as good maternal care as the non-delinquent boys from the same social background, while the serious offenders— and in particular the repeaters—received much less satisfactory care.

Both at the primary and at the secondary schools the teachers reported that the parents of boys who got into trouble showed a low level of interest in their school work, and this was particularly so for the parents of serious offenders and repeaters.

A more factual indication of attitudes is given by the extent to which women used the available maternity and child welfare services at the period the child was born. The parents of the serious offenders, and even more of the repeaters, were less likely to have been to post-natal clinics, attended child-welfare centres or had their children immunized against diphtheria than either the parents of the trivial offenders or of non-deliquent boys from similar families.

It is generally considered that delinquent behaviour is more common among boys who come from large families. Once the social background of the families is taken into account, however, it seems that the trivial and serious offenders do not have more brothers and sisters than would be expected, but that the repeaters do—71 per cent of them come from families of four or more children as compared with 44 per cent of the serious offenders. For the repeaters it is the eldest and middle children who are over-represented, whereas it seems that the youngest children are relatively unlikely to be repeatedly delinquent.

In large families, short birth intervals are equally common among the delinquent and the non-delinquent boys, and from this it seems unlikely that excessive calls on the mother from numerous young children account for low assessments of maternal care and the insufficient use of the maternity and child welfare services which have just been described.

The housing conditions of the delinquent boys were also unsatisfactory, even after taking into account their social background. A high proportion of the serious offenders and repeaters lived in overcrowded homes and many were sharing their beds at the age of eleven and had done so all their lives. In contrast, the trivial offenders were living in housing conditions that were similar to those of the boys who had not been in trouble.

In a previous chapter Dr. Bowlby discussed some of the impressive evidence that links psychopathic behaviour in later life with family disturbances such as death, divorce or separation of

I

the parents and other events that may have resulted in the rupture of affectional bonds or the prevention of their being established. He includes in psychopathic behaviour acts against society, that is to say, crime. It is perhaps an exaggeration to call all the delinquent behaviour we have been discussing in this National Survey ' crime ', but some of the anti-social acts recorded were of a serious nature.

In the National Survey (see Table III) we have contemporary information from a number of sources about home disturbances, some of which will be discussed below. Not all of

TABLE III

SOME CHARACTERISTICS OF FAMILY INSECURITY RELATED TO DELINQUENCY
(Percentage Table)

	TYPE OF DELINQUENCY			
	TRIVIAL	SERIOUS	REPEATERS	EXPECTED*
Father dead	7†	6	4	7
Mother worked (at 11 yrs.)	58	57	62	60
Separation of child in :				
a, unfamiliar surroundings	13	21	23	16
b, familiar surroundings	13	15	18	16
Father frequently away from home	12	9	16	9
Divorce or separation of the parents :				
a, by 6 years	4	12	9	3
b, later	2	4	2	2

* See Note, Table II.
† i.e. the fathers of 7 per cent of the trivial offenders were dead.

these family disturbances were associated with high risks of delinquency, and we start by giving a brief account of the negative findings and then discuss the positive ones in rather more detail.

There are four groups of circumstances occurring in childhood which at first sight would seem likely to be associated with later delinquency but which, once family background is taken into account, are not.

a. There is no excess of delinquents among the sons of men who are unemployed; prolonged unemployment in this sample was almost wholly owing to ill-health.

b. There is no excess of delinquents in those families where the father has died. This was so whether the death had occurred early in childhood or during the school years and whether the death had been sudden or the sequel to a long illness.

c. There is no excess delinquency among children from families in which the mother goes out to work. This statement, however, refers to all the families in which the mother was working—we have yet to look at the question of when she goes out to work and whether she makes satisfactory arrangements for the care of her child when he gets back from school. The information for this closer scrutiny will shortly be available.

d. The separation of a child from his parents for one week or more during the first four years of life is not associated with later delinquent behaviour, so long as the child was not removed from his home environment, or, if removed, so long as he was staying with relations or other persons who were well known to him.

We turn now to the positive findings (see Table III). Separation or divorce of the parents is more frequent among the delinquents—approximately twice as frequent as would be expected after taking into account the home background of the families. This however, is, put into perspective by noting that, even in the families broken by divorce and separation, 77 per cent of the boys were not delinquent. Divorce and separation is associated with serious and repeated delinquent acts rather than with trivial delinquencies, and it is particularly in those families where there has been an early break-up—before the child was six—that delinquent behaviour is found. Later divorce or separation—when the child is at school and when he is of an age to commit delinquent acts—is not significantly associated with higher risks of delinquency.

Early separations from the parents, which involved admission to hospital or other institution, or staying in a strange house with people that were unknown to the child, were associated with an increased risk of later delinquency. This association was with serious and repeated rather than with trivial offences. However, as was emphasized with divorce and separation, the majority of children separated in this way did not show later delinquent behaviour.

After these brief comments on the families of the delinquents some of the characteristics of the delinquent boys themselves, and their behaviour in and out of the classroom, will be discussed (see Table IV).

TABLE IV

SOME CHARACTERISTICS OF THE DELINQUENTS AT SECONDARY SCHOOL
(Percentage Table)

	TYPE OF OFFENCE			
	TRIVIAL	SERIOUS	REPEATERS	EXPECTED*
At non-selective school	60†	76	82	67
Poor worker	34	51	51	28
Troublesome	28	49	56	20
Many absences	57	56	60	40
Known truants	15	27	38	10

* See Note, Table II.
† i.e. 60 per cent of the trivial offenders were at non-selective schools.

The teachers, when these boys were aged thirteen and fifteen, were asked to assess them on a number of dimensions of behaviour in the classroom and their attitude to work. The trivial offenders were not specially picked out by their teachers as being either poor workers or badly behaved, but the serious offenders and repeaters received poor assessments both for work and class behaviour. All the delinquents—trivial, serious and repeaters—were frequently absent from school and sometimes truant, though both truancy and many absences were most often reported for the repeaters. The excess of absences among the delinquents was certainly not owing to illness, for the pattern of sickness among the delinquents was similar to that in the other groups.

The teachers were asked to rate the out-of-class behaviour of these pupils on a variety of items designed to pick out the nervous and the aggressive pupils. On these ratings a relatively large proportion of serious offenders and repeaters were picked out by the teachers as being either aggressive or as having a mixture of nervous and aggressive traits; in contrast, nervous traits alone were less often reported for the delinquents than for the non-delinquents (see Table V). That the teachers were not influenced by known official delinquency was shown by the

TABLE Va

DELINQUENCY RELATED TO TEACHERS' ASSESSMENTS AND NEUROTICISM
(Percentage Table)

SCORES ON NEUROTICISM SCALE

	UPPER THIRD	REST	ALL
Teachers' assessments	% Delinquent	% Delinquent	% Delinquent
Very aggressive	29	18	23
Mixed nervous and aggressive	28	11	19
Very nervous	12	4	6
Rest	15	7	9
All	18	7	11

fact that many of the delinquents were identified by the teachers as being aggressive or having mixed traits before they appeared in the courts.

The pupils, when they were thirteen, filled in a self-rating inventory; this was designed to give three scales, of neuroticism, aggression and intro- or extraversion. The intro/extraversion scale did not distinguish the delinquents from the rest but the scores of the delinquents were high on the aggression and neuroticism scales, particularly the latter. The so-called 'neurotic' statements in this inventory often referred to events that caused anxiety. It will be seen (Table V A) that the boys with a high neurotic score are more likely to be delinquent at every level of the teachers' assessments of aggressiveness and nervousness; thus, only 4 per cent of those regarded by teachers as very nervous were delinquent if they had neuroticism scores that fell into the lower two-thirds, whereas 12 per cent were delinquent if they had neuroticism scores in the top third.

The delinquent boys were reported by their mothers as showing many signs and symptoms of disturbed behaviour, such as bed-wetting, nose-picking and other habits. As, however, the mothers of many of these boys were finding difficulty in coping with their children and their homes, it is probable that the reporting of symptoms was deficient and that the differences given in Table V B are underestimates.

TABLE Vʙ

Delinquency related to Teachers' Assessments and Symptoms
(Percentage Table)

Habit Symptoms reported at 15 yrs.

Teachers' assessments	2 OR MORE	ONE	NONE	ALL
	% Delinquent	% Delinquent	% Delinquent	% Delinquent
Very aggressive	23	26	19	23
Mixed nervous and aggressive	22	22	14	19
Very nervous	9	6	6	6
Rest	10	12	5	9
All	14	12	8	11

In view of the poor attitude and behaviour reports given by their teachers, it is not unexpected to find that the delinquent boys made poor scores in the tests of attainment, which were given to all members of the sample, when they were eight, eleven and fifteen years of age. There is, however, no suggestion that their test performance deteriorated during their years at school; this is of considerable interest because it suggests that the basis of the low attainments of the delinquents lay in attitudes that were acquired before school, or at least in the very early years at school. All the delinquents had lower scores in the verbal than in the non-verbal tests, and their attainment in reading and arithmetic was also poor. The trivial offenders had attainment scores that were only slightly lower than those to be expected from their verbal scores, but the serious, and especially the repeating, offenders had reading and mathematics scores which were considerably below their non-verbal test scores.

TABLE VI

TEST SCORES * OF THE DELINQUENTS : a, AGGREGATE TEST SCORES
AT 3 AGES ; b, INDIVIDUAL TEST SCORES AT 15 YRS.
(Mean Scores)

		TYPE OF OFFENCE			EXPECTED SCORES AFTER STANDARDIZING FOR SOCIAL CLASS
a. Aggregate Test Scores		TRIVIAL	SERIOUS	REPEATERS	
Test	8	46·0	46·8	42·9	49·2
Score	11	47·0	46·4	41·5	49·3
at:	15	46·7	46·6	42·7	50·1
b. Individual Test Scores at 15 yrs.					
Non-verbal		48·4	50·0	46·6	51·2
Verbal		45·7	47·1	43·2	49·2
Mathematics		47·9	46·5	45·0	51·0
Reading		48·0	46·1	42·9	50·4

* All tests have a mean of 50 and a SD of 10.

The delinquents on the whole left school relatively early and very few stayed on past the age of sixteen. At the non-selective schools 6 per cent of the trivial offenders, 5 per cent of the serious and none of the repeaters were at school at sixteen whereas 20 per cent of the non-delinquent boys from similar families stayed until that age or longer. The fact that the trivial offenders left school early was no surprise; for these boys were often absent, frequently truant, and probably were showing an increased discontent with school discipline.

CONCLUSION

To summarize, this chapter describes some of the characteristics of the delinquents and their families as they appear in a National Study.   There are clear associations between delinquency and the educational hopes and aspirations of the parents, and their ability to cope with family problems and to make use of available services both medical and educational.   There is evidence that certain types of family insecurity are associated with later delinquent behaviour, though other types, such as the death of the father, are not.   Delinquent boys were more anxious and at the same time more aggressive than the non-delinquent; in their work at school they were consistently poor scholars, badly behaved in classroom and uninterested in their work at all stages of their school careers.   They left school relatively early and were likely to be absent frequently and also truant.   These remarks apply mainly to the serious offenders and repeaters.   The family circumstances of the trivial offenders, in contrast, were little different from the non-delinquent boys, although during their secondary school years they were fairly frequently absent, truant and left school early.

APPENDIX

SOCIAL CLASS CLASSIFICATION IN THE NATIONAL
SURVEY

The classification is based, for the most part, on the 1957 occupation of the father of the survey child; where this is not known, on the 1946 occupation.

Upper middle class

The father is a non-manual worker, and

a. both parents went to secondary school and were brought up in middle class families, or
b. both parents went to secondary school and one parent was brought up in a middle class family, or
c. both parents were brought up in middle class families and one parent went to secondary school.

Lower middle class

The rest of the non-manual workers' families.

Upper manual working class

The father is a manual worker, and

either the father or mother or both of them had a secondary school education, and/or one or both of them were brought up in a middle class family.

Lower manual working class

The father is a manual worker, and

both the father and the mother had elementary schooling only, and both the father and the mother were brought up in manual working class families.

SEASONAL VARIATION IN HUMAN SEXUAL ACTIVITY

A. S. PARKES

Christ's College, Cambridge

INTRODUCTION

I WAS in some difficulty about the title of this paper. What I want to discuss is seasonal variation in human sexual behaviour as indicated by seasonal variation in the birth rate. Unfortunately, sexual behaviour does not necessarily have, and often does not have, any connection with reproduction. Apart from other manifestations, even intercourse has been said currently to result in pregnancy only once in two thousand times[10]. On the other hand, reproductive activity is too restricted a term in the context of this Symposium. My eventual title, therefore, is a compromise, and like most compromises embodies the worst of both worlds. I turn first to the biological background of my theme.

THE BREEDING SEASON

Many mammals breed only during a restricted period of the year, the so-called breeding season. In the off season, reproductive function, physiological and ethological, is in abeyance in the female and usually in the male. This restriction may be of genetic origin, so that, as in some equatorial animals, it occurs in an environment in which there is no obvious seasonal change, or that it seems to be independent of environmental changes. In either case, the cycle is presumably regulated by a built-in biological clock. More frequently, the incidence of the breeding season is determined by a combination of genetic and environmental factors, in that the animal has a genetic capacity to adjust the breeding season to the environment. In such cases the breeding season, though sharply defined, is susceptible to external influences, notably light. The effect of external factors, however, is not necessarily the same in different species. In the ferret the breeding season is associated with the increasing light of spring, in the sheep with the waning light of late summer and, in both, the season shifts appropriately on transference from the northern to the southern hemisphere, or under conditions of artificially changed lighting. Again, an animal may have the capacity to breed at all seasons of the year, but show some reduction at certain times depending on environmental, nutritional or other conditions; examples of this type of seasonal breeding are to be found among many

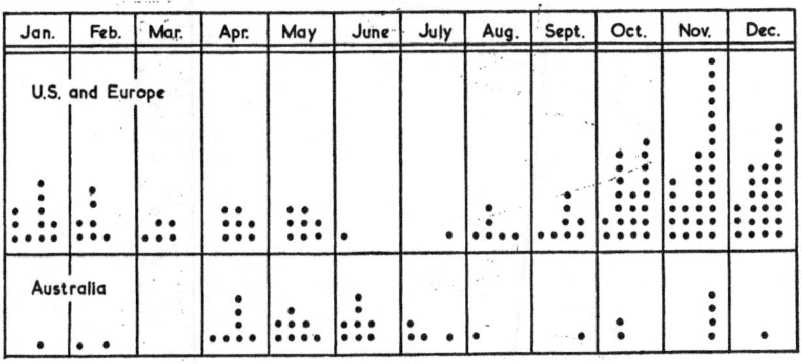

Jan.	Feb.	Mar.	Apr.	May	June	July	Aug.	Sept.	Oct.	Nov.	Dec.

U.S. and Europe

Australia

FIGURE 1

SEASONAL INCIDENCE OF CONCEPTIONS IN RHESUS MONKEYS

After Hartman.[5]

animals. Long ago, Hartman[6] called attention to the incidence of breeding in Rhesus monkeys under laboratory conditions.

Figure 1 shows clearly the concentration of conceptions within the last quarter of the year in the northern hemisphere, and the tendency to reverse this incidence when transferred to the southern hemisphere. There is, of course, more recent information about primates (e.g. Dede and Plentl[4]), but Hartman's data form an appropriate background to my theme.

Against this background, what do we know about man? Reports of seasonal breeding have, of course, come from many parts of the world, but the only noteworthy ones emanate from the Arctic regions, where the alternation of six months' daylight

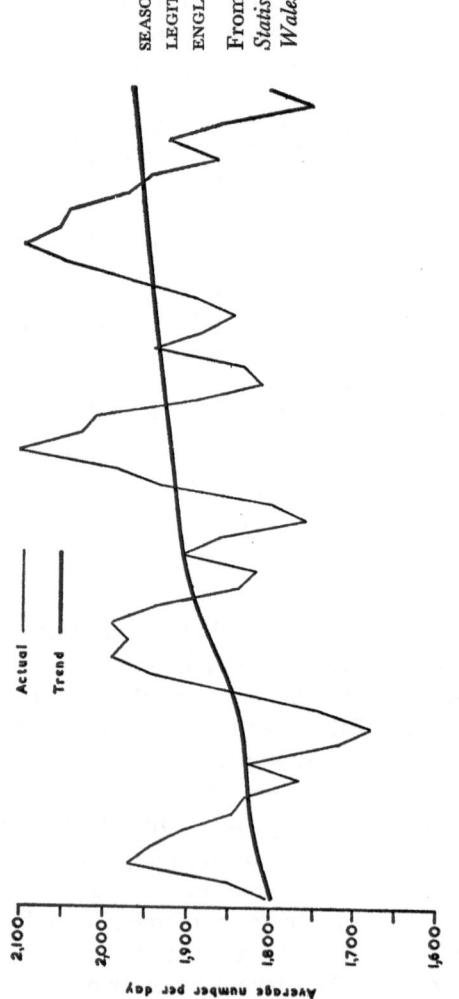

FIGURE 2

SEASONAL VARIATION IN THE LEGITIMATE BIRTH RATE IN ENGLAND AND WALES, 1956–9

From the Registrar General's *Statistical Review of England and Wales*, 1959. Part III. Commentary.

and six months' darkness might be expected, by analogy with animals, to exert a maximal environmental influence. Thus, Dr. Cook, ethnologist to the first Peary North Greenland expedition, writing of the Eskimos there, said: " During the long Arctic night the menstrual function is usually suppressed, not more than one woman in ten menstruating ". And again: " During the whole of this long Arctic night the secretions are diminished and the passions suppressed, resulting in great muscular debility ". Some subsequent observers tended to confirm this report, but contrary ones were not lacking; for instance, Whitaker[13] records that a Danish Governor of a colony in Northern Greenland married a full-blooded Eskimo woman, raised a family and, as a result of his first-hand information, reported that there was no apparent decrease in menstruation or sexual desire during the winter months. More factually, Bertelson[2] analyzed the incidence of 16,101 births in West Greenland between 1901 and 1930. The results indicated that conceptions took place freely throughout the year, but there were two small peaks—in December during the Arctic night and in April on the return of the sun.

SEASONAL VARIATION IN THE BIRTH RATE

With the disappearance of the Eskimo legend, it seems that we have no evidence of a sharply defined and restricted breeding season in man. We do, however, have clear evidence from many parts of the world of repetitive seasonal patterns in the birth rate. We may start at home by considering the situation in England and Wales (EW).

Figure 2 shows, for each of the four years, a high peak of births in March, corresponding to conceptions in June, and a lesser peak in September, corresponding to conceptions in December. The same pattern is seen both in earlier and in later years. Moreover, the phenomenon is not peculiar to England and Wales. Leidl's[8] extensive analysis of the figures for Bavaria shows a well-marked peak of births in February, corresponding to conceptions in May, and a lesser one in September (Figure 3).

The same sort of thing applies to many other parts of the

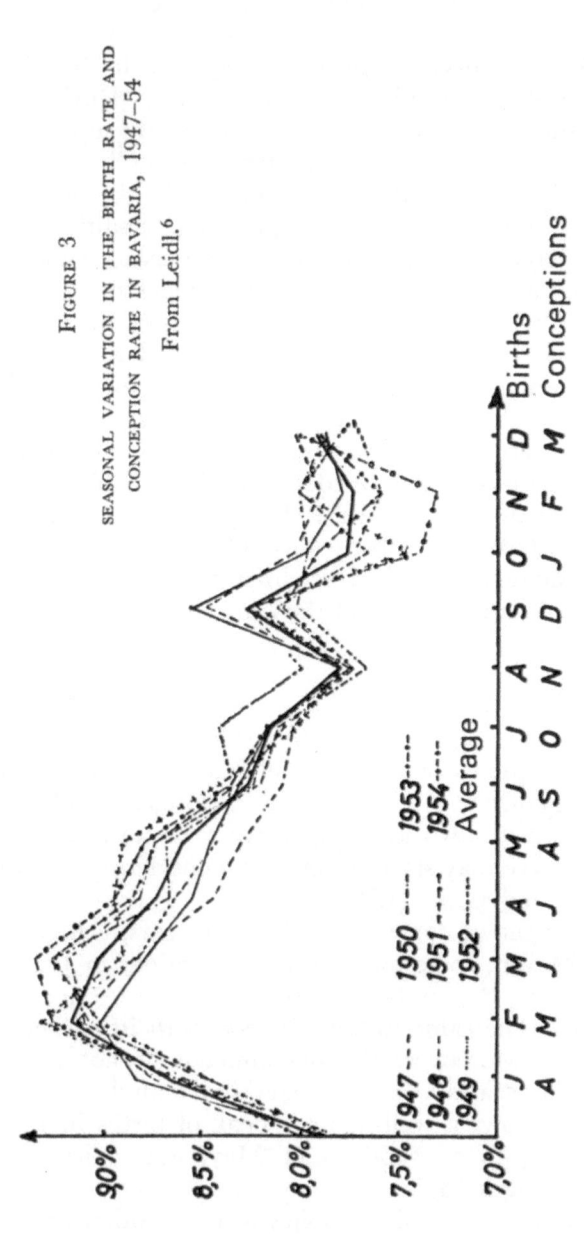

FIGURE 3

SEASONAL VARIATION IN THE BIRTH RATE AND
CONCEPTION RATE IN BAVARIA, 1947–54

From Leidl.[6]

world, and the various patterns have recently been reviewed by Cowgill,[3] who shows, among a wealth of other fascinating material, that the EW pattern is reversed in New Zealand and still more noticably in South Africa, where the white, coloured and Asiatic races are essentially similar in this respect. This reversal in the southern hemisphere is reminiscent of the ferrets, sheep and monkeys already mentioned; but it is difficult to

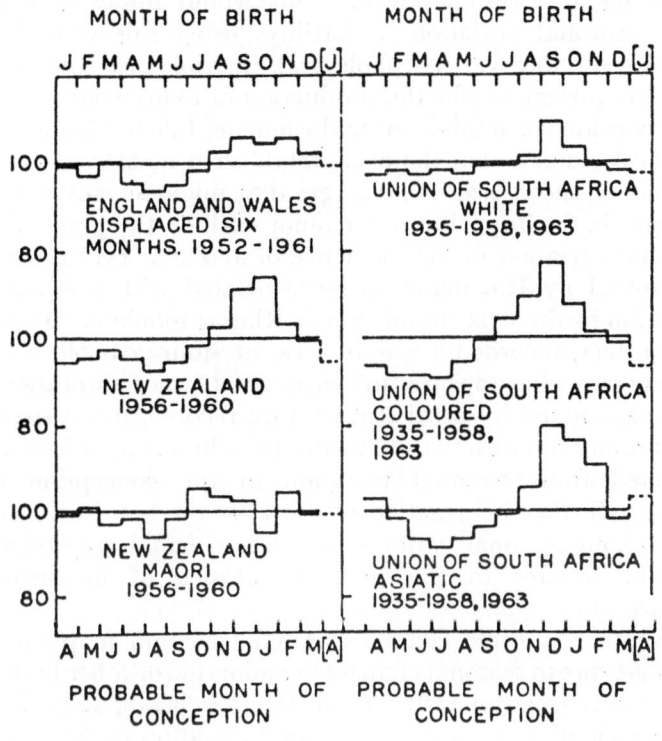

FIGURE 4

REVERSAL OF THE ENGLAND AND WALES PATTERN OF SEASONAL
VARIATION IN THE SOUTHERN HEMISPHERE

From Cowgill.[3]

believe that biological effects are involved, because the EW pattern is reversed also in the United States, where the peak number of births is in September, corresponding, as in South Africa, to conceptions in December (Figure 4).

Considerations such as these raise in acute form the question of what factors are involved.

BIOLOGICAL FACTORS

It has often been said that the seasonal variation in births in modern man is evidence of a primitive breeding season now operating in a vestigial way. This would imply at least a slight seasonal variation in fertility, using this word in its widest sense to cover morphological, physiological and behavioural requirements for the production of living young. Have we any evidence of this ? As to the female, I do not know of any reliable evidence that the overt established menstrual cycle of the female shows seasonal changes that might suggest seasonal changes in fertility; but one cannot exclude the possibility of seasonal variation in the incidence of anovular cycles, such as was found by Hartman [6] to be associated with the seasonal variation in the conception rate in Rhesus monkeys. Döring [5] has, in fact, recorded a case of a twenty-four-year-old woman who showed what appeared to be a seasonal succession of anovular cycles as judged by basal temperature records, but it spanned the summer months, not the winter months as might have been expected from seasonal variation in the conception rate. Pending further information of this kind the best evidence we have about seasonal variation in human female reproductive function relates to the first appearance of menstruation (Figure 5).

Another possible factor—seasonal variation in pregnancy wastage, due to seasonal changes in endocrine or other functions —is also of great interest. According to Belavalgidad [1] there is in New Delhi a very marked seasonal variation in the abortion rate, which is highest in March, April and May, corresponding to conceptions in, say, November, December and January. The birth rate is highest in August, September and October, also corresponding to conceptions in November, December and January. In such a case the variation in the abortion rate augments rather than diminishes the variation in the conception rate calculated from the birth rate. Another aspect of the matter is suggested by the work of Petersen, [9] who found in

Chicago seasonal variation in the conception rate of malformed infants, which was highest in March, April and June, and lowest in August when the conception rate for normal infants was approaching its high point. If the peak conception time of malformed infants indicates also the peak conception time

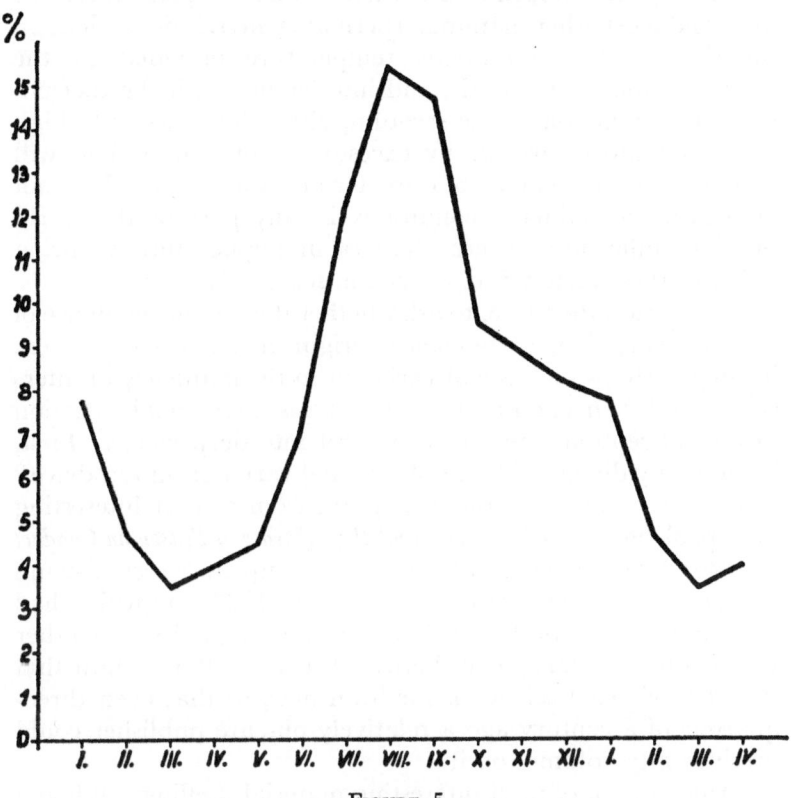

FIGURE 5

SEASONAL VARIATION IN OCCURRENCE OF MENARCHE: THREE MONTHS MOVING AVERAGE

After Valsik.[12]

of early pregnancy wastage, then total conceptions should show less seasonal variation than is suggested by calculating back-wards from the birth rate. There is thus a possible discrepancy between the Indian and US data.

K

So much for the female. What are the biological possibilities of seasonal variation in male fertility? I do not know of any evidence of seasonal variation in the intensity of spermatogenesis, as indicated for instance by sperm counts, of a kind which might imply the existence in the human male of a vestigial breeding season of genetic origin. But there is another possibility. In man and most other mammals spermatogenesis is dependent on the slightly lower-than-body temperature provided by the scrotal position of the testis, and interference with the thermo-regulatory function of the scrotum, either by excessively high ambient temperatures or by excessive scrotal insulation, will depress spermatogenesis to a greater or lesser extent. I do not know whether climatic conditions in any part of the world provide sufficiently severe changes in temperature to bring either of these factors into operation seasonally.

It remains, therefore, to ask whether there is any evidence of seasonal variation, of biological origin, in sex drive in men, leading perhaps to seasonal variation in the frequency of inter-course and thence, on a probability basis, to seasonal variation in the conception rate. Here we get into deep water. First, is there any direct evidence of seasonal variation in sex drive? And here I want to show a diagram from a most interesting little book entitled *Illegitimacy and the Influence of Season on Conduct* by Alfred Leffingwell,[7] which I picked up in a second-hand bookshop in Manchester for ninepence in 1922, and which had originally been published in 1892, together with fifty or so other volumes in a Social Science Series. I mention this to show that the idea of social science is far from new, so that even three-quarters of a century ago a relatively obscure publisher could produce fifty volumes on it.

Among a lot of most interesting material, Leffingwell had a diagram of the seasonal incidence of what he euphemistically called " offences against chastity. " This diagram is reproduced in Figure 6.

The concentration of such offences in the second and third quarters of the year is perhaps suggestive of seasonal variation in sex drive, but it could also be merely the result of variation in opportunity. In any case, the implications are not straight-forward, because another diagram given by Leffingwell shows a similar seasonal variation in the suicide rate, as do the data of Takahashi [11] for another part of the world seventy years later. Moreover, seasonal variation in male sex drive could produce seasonal variation in the birth rate only if the male determines, for the most part, the frequency and timing of intercourse, and if there is a relation between the intercourse rate and the birth rate. The first of these propositions is still probably largely true, and the second must also have been true in the days

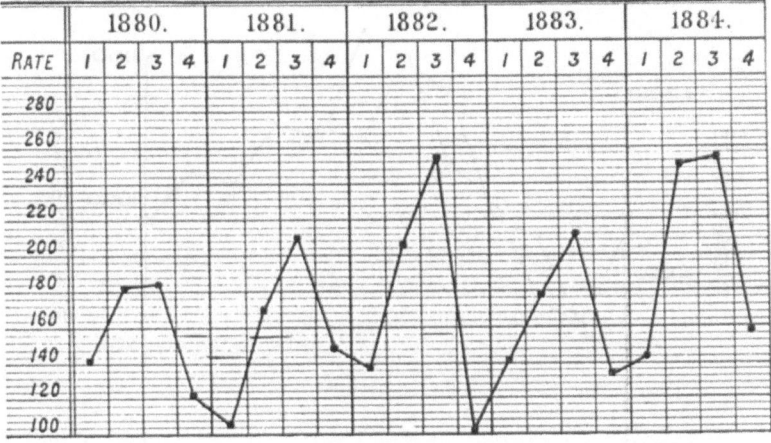

FIGURE 6

SEASONAL INCIDENCE OF OFFENCES AGAINST CHASTITY, 1880–4

From Leffingwell.[7]

before the general use of positive contraception. Even now, the intensity of the urge to intercourse may well affect the willingness to use a cumbersome or frustrating method of preventing conception, but seasonal variation in the frequency of intercourse may nevertheless be greater than the variation in the conception rate.

I do not know of any direct information on this problem, but I want here to record a very interesting and, so far as I know, highly original observation. The Marie Stopes Memorial Centre has a considerable mail-order and over-the-counter trade in contraceptives, mainly in what used to be known, before the pill became respectable, as 'conventional types'. These sales,

analyzed by Miss Faith Schenk, the Centre's secretary, show a
distinct seasonal variation, and one would expect such variation
to be correlated inversely with the conception rate. In fact, the
correlation is a positive one (Figure 7). This unexpected
result, which I propose to refer to as the 'Schenk phenomenon',

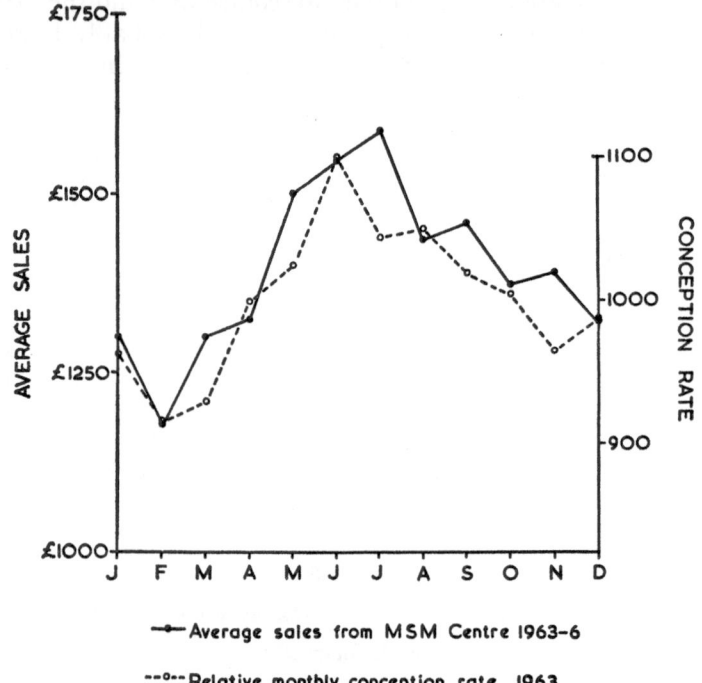

—•—Average sales from MSM Centre 1963-6

--°--Relative monthly conception rate 1963

FIGURE 7

SEASONAL VARIATION IN SALES OF CONTRACEPTIVES BY THE MARIE
STOPES MEMORIAL CENTRE 1963-6, COMPARED WITH SEASONAL
VARIATION IN THE CONCEPTION RATE IN 1963

seems to me inevitably to imply a much increased frequency of
intercourse during the summer. It also poses a question. Is
there seasonal variation in the conception of unwanted children?
Here we have a clear gap in our biosocial knowledge. Although
strict classification is obviously impossible, children may be
thought of as coming into one of three categories: planned,
unplanned but welcome or at least acceptable, unwanted. It

is difficult even to guess the relative proportions of the three categories, but it is pretty certain that the proportions are changing and will change more with the spread of methods of contraception the pill and the IUCD which do not depend on action at the time of intercourse. Even when these are re-inforced by ' morning after ' techniques, which will soon be the case, it is unlikely that unwanted children will disappear com-pletely, but we can hope for a large decrease in their number.

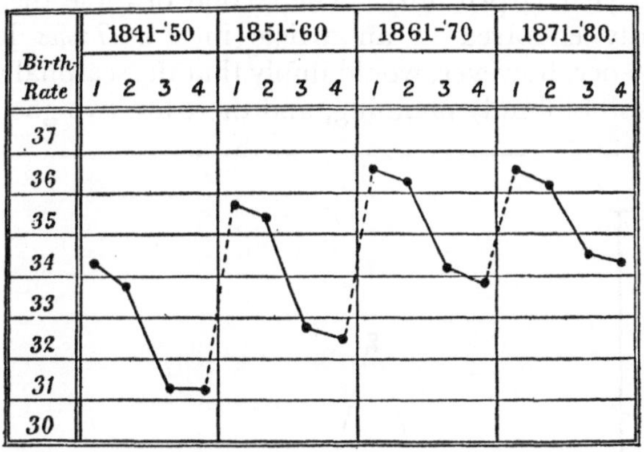

Birth Rate	1841-'50				1851-'60				1861-'70				1871-'80.			
	1	2	3	4	1	2	3	4	1	2	3	4	1	2	3	4
37																
36																
35																
34																
33																
32																
31																
30																

FIGURE 8

SEASONAL VARIATION IN THE BIRTH RATE IN ENGLAND
AND WALES, 1841–80

From Leffingwell.[7]

With them will disappear a fascinating body of knowledge, which can be obtained only before actions have been rationalized or memory blurred by time, of the very human problem of the circumstances attending the conception of the unwanted child: *joie de vivre*, irresponsibility, failure of coitus interruptus or of positive contraception, reaction from a quarrel or other emotional upset, taking a chance, near-rape by the husband, and so on and so on. Before the pattern changes for ever, this slice of biosocial history should be recorded, and an investigation of the why, when and where of the unwanted conception, of the epidemiology of the unwanted child, should be put in hand.

SOCIAL FACTORS

Here, I shall confine myself to what goes on in the UK. In explanation of the spring peak in the birth rate (Figure 2), it could be argued, in the context of the mid-twentieth century, (*a*), that people prefer to have their babies in the spring when the nights are getting shorter and the mornings lighter and/or (*b*), that to get the maximum tax rebate for the minimum of expense, people arrange to have their babies near the end of the fiscal year (April 5th). The second of these theories was confidently endorsed in an article in *The Times* last year. Either theory, however, would imply that the seasonal variation is the result of family planning, and there are two good reasons

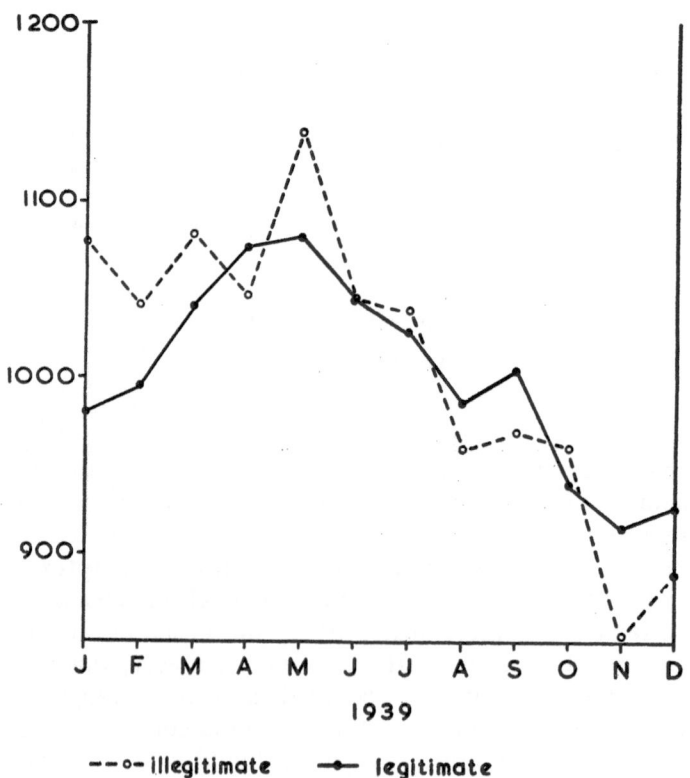

1939

--o-- illegitimate   --•-- legitimate

for rejecting this implication. In the first place, the same phenomenon occurred a century ago, long before the days of income tax and when family planning, if any, was rudimentary (Figure 8). In the second place, the same cycle is seen with illegitimate births, few of which, presumably, are planned. In 1939 the seasonal variation in illegitimate births was even greater than in legitimate ones, which is what one might expect if the majority of illegitimate conceptions took place under what Marie Stopes euphemistically referred to as ' alfresco '

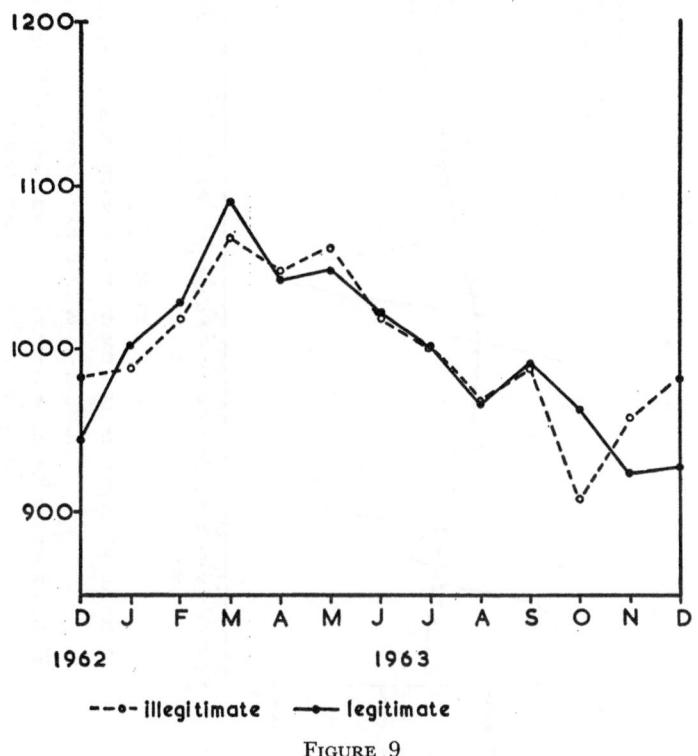

FIGURE 9

COMPARISON OF SEASONAL VARIATION IN LEGITIMATE AND ILLEGITIMATE
BIRTH RATE IN ENGLAND AND WALES.

a, 1939 (opposite page) ; *b*, 1963

Plotted from the Registrar General's *Statistical Review of England and Wales*, 1939; 1963. Part III, Commentary.

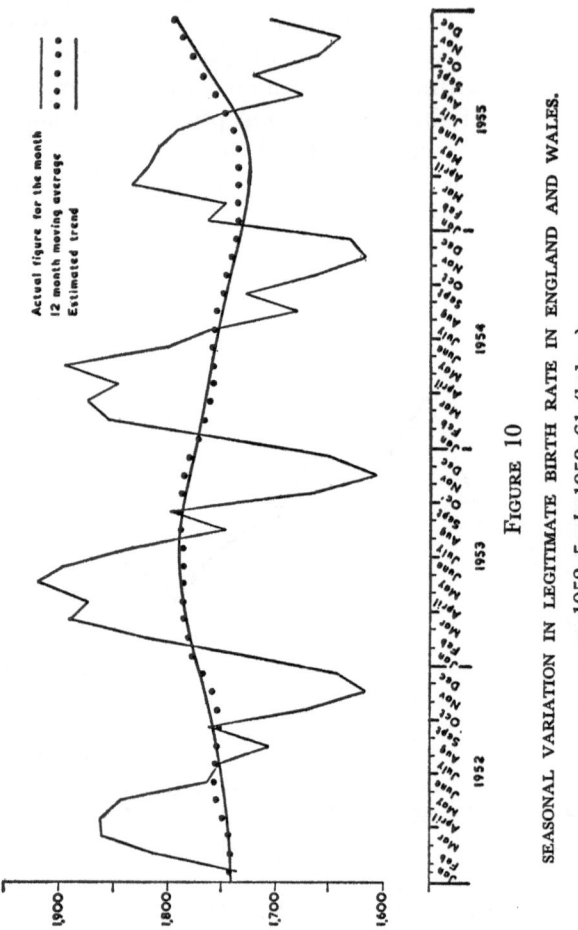

FIGURE 10

SEASONAL VARIATION IN LEGITIMATE BIRTH RATE IN ENGLAND AND WALES.

a, 1952–5; *b*, 1958–61 (below)

Plotted from the Registrar General's *Statistical Review of England and Wales*, 1955; 1961. Part III, Commentary.

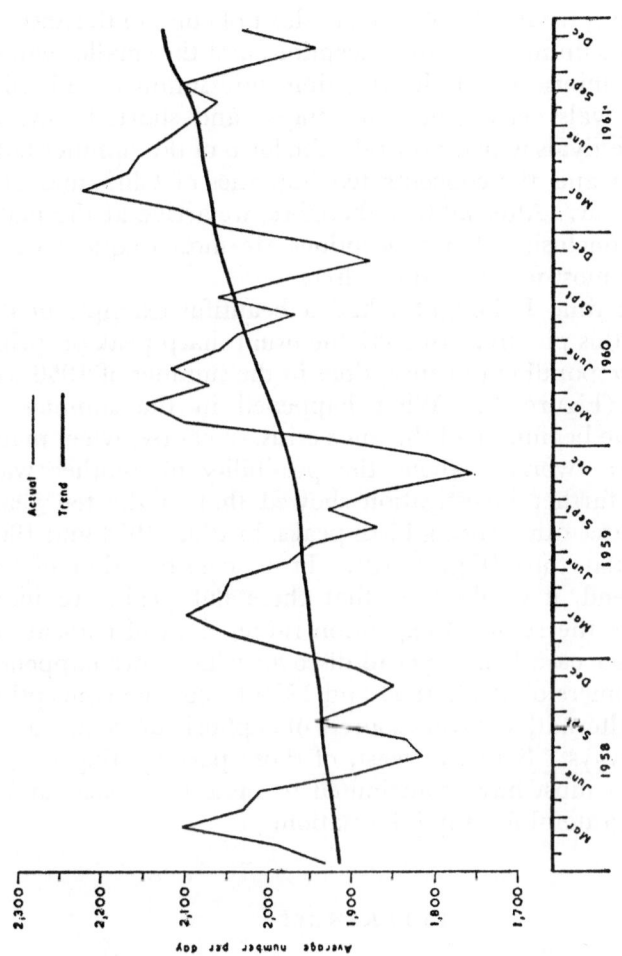

conditions. This difference may now have disappeared, which in itself is a significant comment on social changes, but seasonal variation still continues in both (Figure 9).

If, therefore, family planning is not the explanation of the annual cycle, what is? We are not short of clues to the answer. The larger summer peak in conceptions and the smaller winter one are reminiscent of the one-time mid-summer and mid-winter festivals celebrating the longest and shortest days of the year, festivals which now take the form of the summer holiday season and the concentrated festivities of Christmas and the New Year. After all this, therefore, we arrive at the platitudinous conclusion that conceptions are most frequent when people are most festive and carefree.

At one time I thought I had a beautiful example of the reverse of this picture. In 1957 the usual sharp peak of spring births corresponding to conceptions in the summer of 1956 was truncated (Figure 2). What happened in the summer of 1956? The beginning of the Suez crisis, of course, when many people were worried about the possibility of another war. However, further investigation showed that in the ten years 1952–61 three other spring birth peaks, in 1953, 1954 and 1960 had had flat tops (Figure 10). From consideration of the general trend, it would seem that these flat peaks are more likely to be the result of expansion rather than of truncation, but in either case I have yet to discover what crises happened in the summers of 1952, 1953 and 1959 to cut the conception peaks, or alternatively what sources of euphoria arose to extend them sideways. Some, at least, of those participating in this Symposium must have contributed to these flat peaks, and I would be grateful for any information.

REFERENCES

1. BELAVALGIDAD, M. I.   1963.   Trend lines and seasonal variations in births and abortions.   *J. Obstet. Gynaec., India* **13**, 23.
2. BERTELSON, A.   1935.   *Meddr Grønland* **117**, No. 1.
3. COWGILL, U. M.   1966.   Season of birth in man.   *Ecology* **47**, 614.
4. DEDE, J. A. and PLENTL, A. A.   1966.   Induced ovulation and artificial insemination in a Rhesus colony.   *Fertil. Steril.* **17**, 757.

5. Döring, G. K. 1958. Über ungewöhnliche Basaltemperaturkurven. *Geburtsh. Frauenheilk.* **9**, 1124.

6. Hartman, C. G. 1932. Studies in the reproduction of the monkey *Macacus (Pithecus) rhesus*, with special reference to menstruation and pregnancy. *Contr. Embryol. Carneg. Instn.* **23**, 1.

7. Leffingwell, A. 1892. *Illegitimacy and the Influence of Season on Conduct.* London. Swan Sonnenschein.

8. Leidl, W. 1958. *Climate and Sexual Function in Male Domestic Animals.* Hanover. Schaper.

9. Petersen, W. F. 1934. The seasonal trend in the conception of malformations. *Am. J. Obstet. Gynec.* **28**, 443.

10. Swyer, G. I. M. 1967. Fertility control : achievements and prospects. The Tenth Oliver Bird Lecture. *J. Reprod. Fert.* **15**, 295.

11. Takahashi, E. 1964. Seasonal variation of conception and suicide. *Tohoku J. exp. Med.* **84**, 215.

12. Valsik, J. A. 1965. The seasonal rhythm of menarche : A review. *Hum. Biol.* **37**, 75.

13. Whitaker, W. L. 1938. The question of a seasonal sterility among the Eskimos. *Science* **88**, 214.

PRIMARY MENTAL ABILITIES OR GENERAL INTELLIGENCE? EVIDENCE FROM TWIN STUDIES

STEVEN G. VANDENBERG

University of Colorado

I FEEL deeply honoured to have been invited to contribute to this Symposium. In my University lectures on individual differences, I always devote time to a discussion of Galton. I do not know whether Galton personally started the Eugenics Society; but at any rate he was a direct forerunner of those of us who are interested in the study of hereditary factors in human behaviour and their evolutionary history. This is true, even though our ideas about innate superiority and the importance of the environment have changed drastically in the last fifty years.

I want to acknowledge my debt to many friends and colleagues who repeatedly have allowed me to use their data, and to my assistants and co-workers. I am grateful to the US National Institutes of Health and the National Science Foundation for support of my research. All the serological tests of the twins' blood have been performed by Mrs. Jane Swanson of the Minneapolis War Memorial Blood Bank. Table I shows the tests we use, which provide 95 per cent

TABLE I

GENETIC MARKERS USED TO DETERMINE ZYGOSITY

ALWAYS USED	NOT ALWAYS USED	
A_1A_2BO	Mt^a	Martin
MNSs	Mi^a	Miltenberger
Rhesus tests CcDEe (Rh factor)	P_1	
Lutheran a and b	C^w	
Lewis a and b	Wr^a	Wright
Kell K	Vw	Verweyst
Cellano k	Yt	Cartwright
Kidd (Jk^a)	Do^a	Dombrock
Duffy (Fy^a)		

accuracy in diagnosing zygosity, which is sufficient, in view of the fact that many of the measures we study are considerably less accurate.

Before we delve into the topic, we may avoid unnecessary disagreement if we state clearly at the outset some issues that will *not* be discussed. We will talk about *hereditary* components of *human abilities*. In other words, there will be no reference in this paper to animal research and what it shows about behaviour genetics, nor to human personality, even though personality affects ability measures enormously, as, for instance, Professor Eysenck [5] has shown for introversion-extraversion and performance on learning tasks.

Most importantly I will not enter into the old controversy, as to whether it is more truthful or more useful to speak of intelligence in terms of a general factor plus some specific factors found in tests not measuring this general factor well, or whether the view is more apt that there are a number of separate abilities, some of which are perhaps somewhat correlated.

I suspect that most British psychologists hold the first view and many American ones the second, because of differences in the homogeneity of the samples of persons studied.

In the United States one encounters a wide variety of students and some of them may have reached the University in spite of marked shortcomings in certain of their high school studies. It is my impression that in the United Kingdom, as on the Continent, more than in the USA, most high school pupils bound for college or university are required to meet certain standards in all subjects before they are allowed to pass on. This alone could account for the controversy. The need for more cross-cultural validation studies happens to be one of my pet themes. A few studies in the United Kingdom and in the United States on similar subjects, using the same test battery, might lead to disappearance of the controversy. In any case, it will eventually be the power to predict success in some endeavour that will be the crucial test of the social usefulness of the competing views. Unfortunately, there have been few studies of Guilford's factors, which form the most extreme expression of the multi-factorial theory of intelligence.

Now that the issues that will *not* be considered are stated, what will I discuss? Regardless of the position one takes in the controversy between those who favour a general factor in intelligence and those who favour the multi-factor views, one can be uncommitted about the dimensions of the hereditary contribution to ability measures.

This is the question in which I have become interested. Are there specific hereditary disabilities or talents? It is generally accepted that the inheritance of intelligence must be controlled by a large number of loci, but how many we do not know. Several facts lead us to expect many. First of all there is the fact that the IQ seems to follow closely a normal distribution. This strongly suggests that many loci are involved, although Professor Thoday has shown that normal distributions may occur even when the number of genes segregating is quite limited.[10] Secondly, we know of a large number of genetic anomalies which all affect intelligence adversely. This alone is proof that many genes are involved in the normal development of intelligence. Could it be that the same or similar genes also control the *variations* in ability levels in the normal range? With perhaps the greater ability the fewer harmful genes? Or are there separate loci for superior ability? Inherent in this idea is the view of intelligence as a unitary trait and of the more or less equal value of the genes controlling intelligence.

What if certain abilities were dependent in part on one or a few alleles at separate loci? One can readily think of some extreme examples of disabilities which do not really fall within the problem under discussion, but which help to focus our thinking. For instance, hereditary deafness would prevent development of musical skills, a crippling condition of athletic prowess, and blindness the appearance of certain mathematical or artistic skills. But what about less drastic limitations, and what of special gifts?

Could there be separate hereditary factors in ability, each due to one or more loci, for language skills such as reading, spelling and the learning of foreign languages, or for arithmetic, algebra and some other forms of mathematics, or for the ability to visualize two- and three-dimensional patterns? I believe that this possibility is subscribed to in a vague sort of way by many of us as the correct view.

We can approach this problem from two directions : one method is rather time-consuming, the other is a more economical short-cut. The first method consists of locating clear examples of persons deficient in only one of the basic skills such as reading,

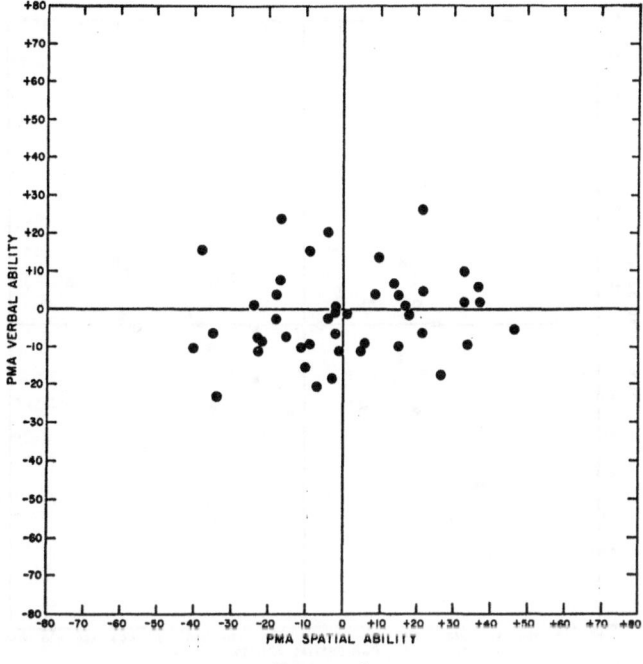

FIGURE 1

CORRELATION BETWEEN MZ TWIN DIFFERENCES IN SPATIAL AND
IN VERBAL ABILITY. $N = 45$ $r = \cdot137$

arithmetic or spatial visualization and collecting information about their relatives to see if this is a genetic deficiency and whether it is specific or whether it affects other abilities. One might even start with individuals with a specific syndrome such as Down's, Klinefelter's, Turner's, etc. If these individuals do indeed show specific psychological deficiencies, we would be able to locate the responsible gene on a specific chromosome.

This clinical approach is of course the preferred method in human genetics. It presupposes, however, that we know fairly well what to look for. It also requires that post-traumatic brain damage can be ruled out, something of which we cannot always be sure.

The second method, which *I* have mainly used, is a modifica-

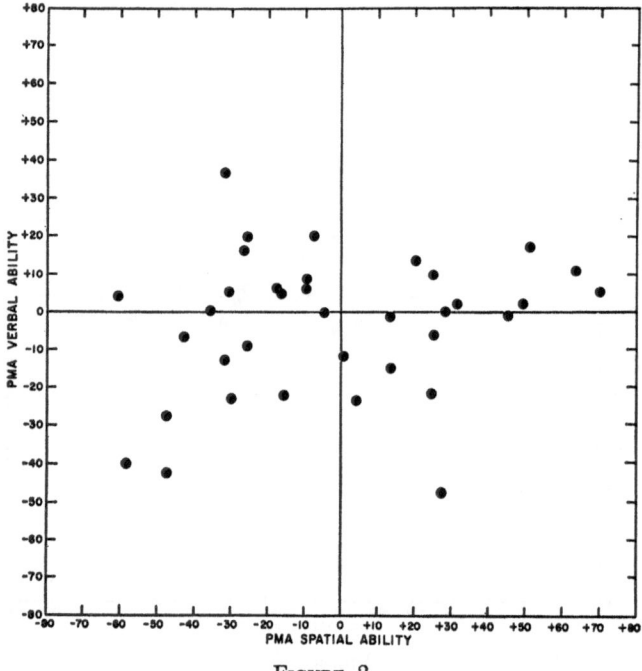

FIGURE 2

CORRELATION BETWEEN DZ TWIN DIFFERENCES IN SPATIAL AND IN VERBAL ABILITY. $N=37$ $r=\cdot215$

tion and expansion of the traditional twin method to the multivariate case.

In essence we are going to ask whether the twin John does better than his brother James on *all* ability tests, as he would if there is one set of genes that affects all abilities, assuming of course that specific environmental influences on different abilities are not too important. Figures 1 and 2 show that the same twin does not always score higher on various tests. If he did, there would only be entries around the leading diagonal, i.e. in the lower and left upper right quadrants. Actually, we have to be more precise. We know now that there are two kinds of twins. Identical pairs are the result of the splitting of one fertilized egg and therefore have the same genes, while fraternal pairs result from the fertilization of two separate eggs by separate spermatozoa, so that half of their genes, on the average, will be different.

Differences between identical twins reared together will not tell us much, except about intrafamily influences. Fraternal twins *will* give us some information. In fact, if we can assume that the influences within the family are on the average equally important in producing identical and fraternal twin differences, we can take a step further. This step consists of comparing the identical and fraternal twin differences, or rather the variances of these two sets of differences, as well as the covariances of differences on different abilities. In other words, we can make a generalization to the multivariate case of the traditional evaluation for evidence of heredity of a variable by the twin method. This means that all the statistical techniques used in twin studies of a single trait can be considered for application in the multivariate situation.

I mentioned earlier the assumption of similar within-family conditions for identical and fraternal twins. I have elsewhere reviewed the objections to this assumption [13] and the sparse evidence and have concluded that the assumption may be more valid than has been supposed by romantic theories of special twin interactions. These, I believe, are derived from biased psychiatric material which is really quite rare and unrepresentative.

In choosing a statistical method, I have preferred the one which has a test of statistical significance. This method is less easily misinterpreted, and also can be generalized to the multivariate case in a more rigorous way. I am speaking of the F test between the fraternal and identical within-pair variance. In addition this method allows further partitioning of the variance, for instance, between sexes, between trials, between alternate versions of the questionnaire, or ability measure, and

L

so on. It will be convenient to use the conventional abbrevia-
tions: MZ for monozygotic or one-egg (identical) twins and
DZ for dizygotic or two-egg (fraternal) twins. Using these
symbols, the formula for the F test is:

$$F = \sigma_W^2 \, DZ / \sigma_W^2 \, MZ, \tag{1}$$

where the within-pair variance of the twins

$$\sigma_W^2 = 1/N \sum_{i=1}^{N} (X_{iA} - X_{iB})^2 \tag{2}$$

and where A and B are twins.

If we rewrite equation (1)

$$\sigma_W^2 \, DZ - F\sigma_W^2 \, MZ = 0, \tag{3}$$

we can generalize this to

$$[\, C_{DZ} - \Lambda \, C_{MZ} \,] = 0, \tag{4}$$

where C is the within-pair variance-covariance matrix, or,
more simply, the matrix of the cross products of the twin
differences on the different variables divided by N.

A typical element of C would be

$$c_{XY} = 1/N \sum (X_{iA} - X_{iB}) \, (Y_{iA} - Y_{iB}) \tag{5}$$

Equation (4) is a canonical equation for which, fortunately,
computer routines are available. There are several ways of
interpreting this equation. Let me only say here that we find
first *that* linear combination of the variables which leads to the
best discrimination of MZ and DZ twins. This will be one
hereditary component. After removing this composite we ask
whether another linear combination still allows us to separate
the two kinds of twins. This would be another independent
hereditary component; and so we continue, as long as the
significance test indicates that another root of the equation
may be interpreted.

Originally [12] I used a significance test for the homogeneity
of the remaining r-k roots after extraction of the first k roots.[1]
This is a test for symmetric matrices only, but I knew of no better
procedure. Since that time Professor Bock has called to my
attention another significance test designed by Bartlett [2] which
is the proper one to use in this situation.[3]

The first time this multivariate analysis was applied was with data on high school students from the Michigan twin study on 45 pairs of identical and 37 pairs of fraternal twins. The variables were the six scores of Thurstone's Primary Mental Abilities test: Verbal, Number, Spatial Visualization, Word fluency, Reasoning and Memory. The first four of these six abilities had a significant hereditary component.[11] Was this hereditary component the same for these tests and the the lack of correlation between the tests due only to environmental influences? The solution of the canonical equation is shown in Table II. Of the six roots, three were significant, even for this rather small sample. If a rotation of the axes

TABLE II

Solution of $| C_{DZ} - \Lambda C_{MZ} | = 0$ for the 6 PMA Scores of 37 fraternal (DZ) and 45 identical (MZ) Twin pairs of High School Age [12]

	I	II	III	IV	V	VI
root λ,	4·000	2·232	1·587	1·001	0·646	0·382
N	0·035	−·030	0·345	−·181	−·112	−·173
V	−·183	0·701	0·186	−·239	−·288	0·207
S	−·219	−·011	0·346	0·029	0·318	0·058
W	0·004	−·401	0·237	0·289	−·547	0·048
R	0·504	0·385	−·099	0·584	0·505	−·248
M	0·815	−·436	0·809	−·696	0·500	0·927

were performed, one might be able to find a rational interpretation of the roots. I have not done so because a rotation is not meaningful with so few variables.

In a Finnish study of adult male twins over forty years of age, eight variables were used.[9] While the actual tests were different, they were quite similar to the ones in the previous study. There was a measure of verbal ability, one of word fluency, two of spatial visualization, two of number ability, and two of memory. The solution of the canonical equation is shown in Table III, where the variables are identified by their initial letter. Four of the roots are significant for this sample. It is tempting to conclude, as I have done sometimes, that the same hereditary components were found in both studies.

TABLE III

SOLUTION OF $\mid C_{DZ} - \Lambda C_{MZ} \mid = 0$ for 8 ABILITY TEST SCORES OF 157 FRATERNAL (DZ) AND 189 IDENTICAL (MZ) ADULT, MALE, FINNISH TWIN PAIRS [9]

V	330*	421	-470	-029	-210	408	-718	-002
W	518	100	080	-877	461	-054	224	-220
S_1	432	414	152	405	284	272	262	-030
S_2	196	281	347	-032	-233	-633	-029	256
N_1	256	-432	-332	008	-476	259	493	601
N_2	-148	-012	-652	184	322	-521	020	-217
M_1	493	-189	120	129	-380	-056	137	-666
M_2	261	-582	282	122	367	-117	-322	182
Size of root	3·556	2·256	1·785	1·682	1·246	1·166	1·132	0·974
H	0·72	0·56	0·44	0·41	0·20	0·14	0·12	0·03

* Decimals omitted.

The next study was by Bock and Vandenberg.[3] This time the variables were the sub-tests of the Differential Aptitudes Test (DAT for short). These scores are not quite so independent of one another as are the PMA sub-tests, but they have higher reliabilities and are extremely well validated against school criteria. The eight scores are Spatial visualization, Numerical reasoning, Abstract reasoning, Verbal reasoning, Mechanical reasoning, Clerical speed and accuracy, Spelling and finally Grammar. The solution of the canonical equation for these variables is shown in Table IV for the boys and in

TABLE IV

DAT ANALYSIS : CHARACTERISTIC ROOTS AND VECTORS OF $\mid C_{DZ} - \Lambda C_{MZ} \mid = O$ FOR BOYS [3]

	VECTORS				
TEST	1	2	3	4	5
1. Spatial	0·245	0·470	0·080	-·473	0·447
2. Numerical	0·304	-·402	0·100	0·214	0·766
3. Abstract	0·401	0·139	-·650	0·054	0·035
4. Verbal	0·500	-·177	0·177	-·022	-·167
5. Mechanical	-·131	0·590	-·195	0·015	0·166
6. Clerical	0·196	0·399	0·191	0·822	-·041
7. Spelling	0·484	-·097	-·339	-·077	-·325
8. Sentences	0·383	0·223	0·583	-·214	-·223
Root	3·529	2·570	0·900	0·623	0·376

Table V for the girls. For the boys there were three significant roots, but for the girls only two. This may be partly due to the relative lack in the girls of spatial visualization and of mechanical reasoning.

TABLE V

DAT ANALYSIS : CHARACTERISTIC ROOTS AND VECTORS OF
$| C_{DZ} - \Lambda C_{MZ} | \%$ O FOR GIRLS [3]

TEST	VECTORS				
	1	2	3	4	5
1. Spatial	0·300	0·468	0·204	− ·542	0·430
2. Numerical	0·434	− ·092	0·054	0·293	0·095
3. Abstract	0·364	0·339	0·091	0·571	0·069
4. Verbal	0·414	− ·194	− ·207	0·048	− ·509
5. Mechanical	0·296	0·562	− ·115	− ·129	− ·427
6. Clerical	0·234	− ·316	0·859	− ·113	− ·188
7. Spelling	0·389	− ·296	− ·218	0·122	0·561
8. Sentences	0·353	− ·341	− ·324	− ·497	− ·092
Root	5·005	1·619	0·747	0·444	0·185

Dr. Bock took the analysis a step farther and constructed the estimated correlation matrix for the heritable part of the DZ twin differences. This allowed an interpretation of the nature of the dimensions. A strong general hereditary component appeared in both male and female samples, plus a second hereditary component defined by the contrast between mechanical reasoning + spatial relations versus numerical + grammar. The third component for the boys was defined by the clerical speed and grammar tests.

Rather than discuss the details of each of these three studies, I should like to propose a general conclusion. Twin studies are not different from other studies: depending on the degree to which the variables in each study are measuring basically different abilities or rather more related abilities, one will find a general hereditary ability factor of low, moderate or high importance plus one or more factors that tend to be limited to one or more abilities. As this canonical method is applied to further test batteries, I expect that we shall find several classes of tests, or rather of abilities. In some tests the hereditary component will be quite high. By checking twin concordance item by item one could increase the hereditary component perhaps, as Loehlin [8] suggested. These tests may in part prove to have some overlapping hereditary variance; in part, the hereditary variance will be independent. Other tests may turn out to have insignificant hereditary components, except in extremely large samples.

It should be remembered that the hereditary component in a test will vary as a function of the age, sex and other aspects of the sample studied. In addition, we have to remember that the tests have to be reliable enough to permit fairly high MZ concordance. Some of the most interesting twin studies in the past used tests which do not meet this criterion.

In proposing that there are several hereditary factors, I am going beyond the idea of Professors Cattell [4] and Horn, [6] who suggested that there are two general ability factors—one largely acquired, the other largely hereditary. Because the acquired ability factor shows up mostly in tests requiring long-term familiarity with cultural conventions, it was labelled by Cattell and Horn *crystallized*, and the hereditary factor *fluid*. At first sight this choice of labels might be thought to be confusing, especially if one were making the mistake of thinking of innate abilities as levels of individual performance fixed at the moment of conception. Of course this is a mistake *we* will not make, but some will. What *is* fixed is the reaction norm or range within which a given person's performance can lie, the environment determining the level actually realized. Because the term *fluid* for the hereditary ability factor makes it impossible to think of the hereditary factor as leading to a fixed level, the term may have pedagogic utility. Nevertheless, the term *crystallized* seems an unhappy choice for an ability factor that is primarily acquired. Cattell and Horn used a conventional factor analysis of data from a number of unrelated children to support their theory. Of course one cannot obtain information about heredity from such a study.

I have already mentioned further multivariate analyses of twin differences that are planned. Until these are done we cannot say to what extent the various abilities found to have a significant hereditary component are based on the same or different sets of genes. As part of a review of our present knowledge I have been collecting the results from a number of twin studies. Using a very crude method—namely simply averaging the F ratios from ten studies and forcing all the tests used in these studies into one of seven categories I came up with the results shown in Table VI.

TABLE VI

COMBINED RESULTS OF 10 TWIN STUDIES SHOWING THE RELATIVE IMPORTANCE OF HEREDITY FOR 7 ABILITIES

ABILITY	F
Verbal	2·53
Word fluency	2·47
Perceptual speed	2·26
Spatial	2·25
Memory	2·18
Number ability	1·91
Reasoning	1·65

The results for the reasoning tests seem a little puzzling. They seem to indicate that heredity plays only a minor role in reasoning, even though this might be regarded as close to the heart of intelligence, if we are to believe Spearman. I would suggest that the reason for the rather low degree of hereditary control is due to an artefact. Many reasoning tests allow one to reach the correct conclusion through several different approaches. Identical twins frequently select different routes, one of which may be speedier. Such more or less accidental differences in the method used would lower the MZ concordance on that test and, hence, the F ratio or whatever statistic is used to evaluate the hereditary component.

Besides the refinements of statistical and pyschological techniques employed in twin studies, future research in human behaviour genetics in the next few years may be expected to test specific hypotheses, perhaps including modelling of various traits under the control of x number of genes with one or more loci contributing more to a trait than the other loci. Perhaps a family study of height may be worth while to-day.

We may also expect more detailed psychological studies of persons with genetic anomalies and of their close relatives.

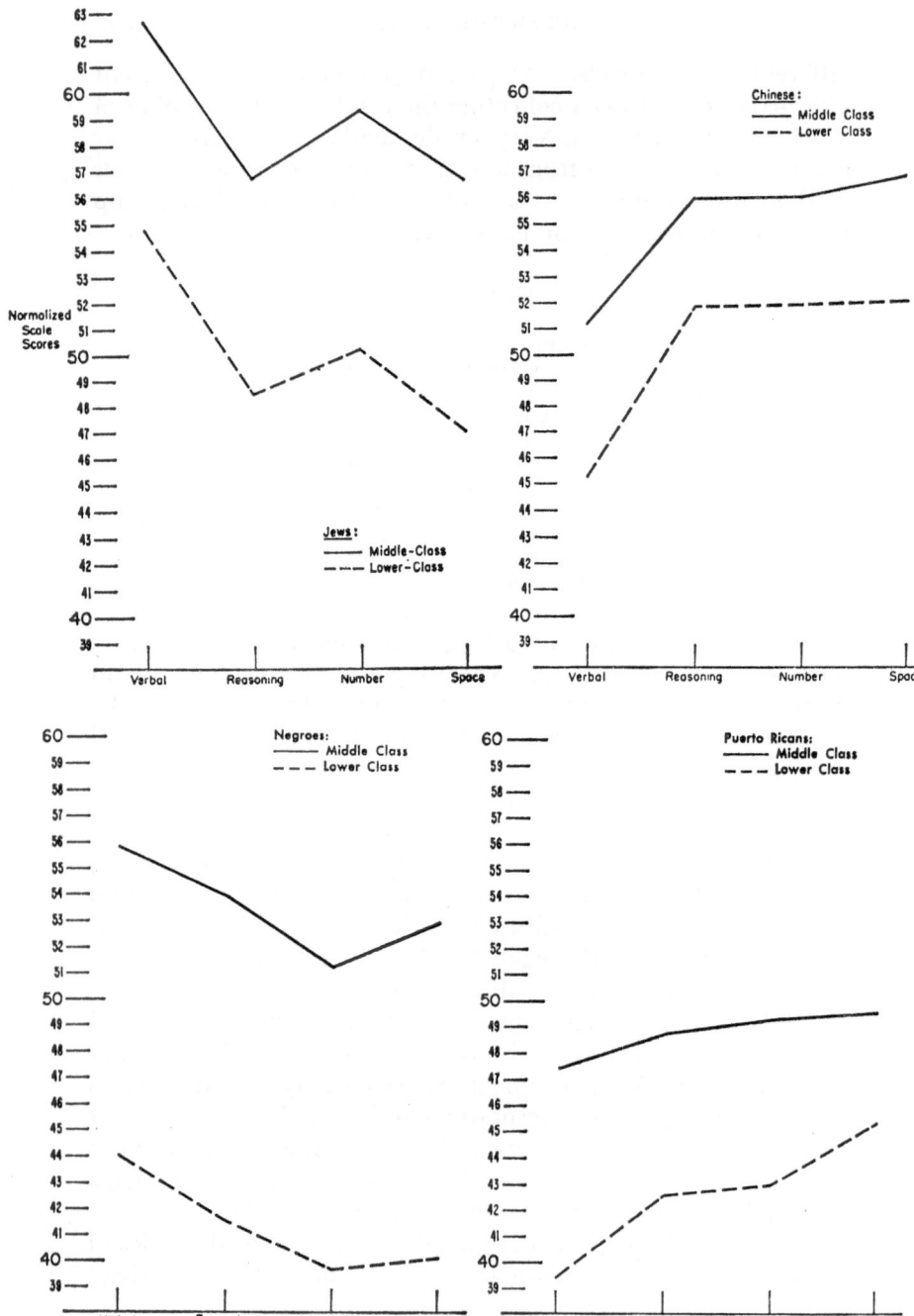

Perhaps someone will start collecting data on children whose personality is markedly different from that of both parents and of their other children, to see whether the same personality can be found among other relatives. I am thinking of the occasional shy child in a sociable, outgoing family or the boisterous, happy child in a subdued, withdrawn family.

If independent hereditary abilities are found and identified it may perhaps be possible to return to a study of national or racial differences in ability with more hope of success, because patterns of ability might be found independent of the level of general intelligence. This possibility is strongly suggested by a study of children from middle and lower class families of four different ethnic groups.[7] Figure 3 shows that in each group the same patterning of abilities occurred for the two socioeconomic classes, even though there were marked differences between the groups. If, as I have suggested, there may prove to be a number of independent hereditary abilities as well as a general hereditary ability factor, one might ask why the situation is so similar for the hereditary components and the phenotypic variables. The answer is, of course, that man's heredity helped to shape his environment and man's environment helped to select his genes, so that it is to be expected that they will fit together rather nicely.

One may ask where these peculiarly human abilities came from. Were they newly invented during man's evolution? I believe not. Many human abilities are present in rudimentary forms in various animals. We find an ability to orient oneself in space in many animals; a primitive number sense exists in some animals; an ability to manipulate objects has been observed in some animals, and so on. In man, these abilities have been perfected and magnified by verbal representation and mental rehearsal, while speech also allowed systematic instruction of the young by parents and by individuals with special skills.

FIGURE 3

COMPARISON OF TEST PERFORMANCE OF CHILDREN FROM MIDDLE AND LOWER
CLASS FOR FOUR DIFFERENT CULTURAL GROUPS [7]

REFERENCES

1. BARTLETT, M. S. 1950. Tests of significance in factor analysis. *Br. J. statist. Psychol.* **3,** 77.
2. BARTLETT, M. S. 1951. The goodness of fit of a single hypothetical discriminant function in the case of several groups. *Ann. Eugen.* **16,** 199.
3. BOCK, R. D. and VANDENBERG, S. G. 1968. Components of heritable variation in mental test scores. In *Progress in Human Behavior Genetics.* Ed. S. G. Vandenberg. Baltimore, Md. Johns Hopkins University Press.
4. CATTELL, R. B. 1967. The theory of fluid and crystallized general intelligence checked at the 5–6 year old level. *Br. J. educ. Psychol.* **37,** 209.
5. EYSENCK, H. J. 1967. *The Biological Basis of Personality.* Springfield, Ill. C. C. Thomas.
6. HORN, J. L. and CATTELL, R. B. 1966. Refinements and test of the theory of fluid and crystallized general intelligence. *J. educ. Psychol.* **37,** 209.
7. LESSER, G. S., FIFER, G. and CLARK, D. H. 1965. Mental abilities of children from different social class and cultural groups. *Monogr. Soc. Res. Child Dev.* **30,** 4. (Whole number 102.)
8. LOEHLIN, J. C. 1965. A heredity-environment analysis of personality inventory data. In *Methods and Goals in Human Behavior Genetics.* Ed. S. G. Vandenberg. New York. Academic Press.
9. PARTANEN, J., BRUNN, K. and MARKKANEN, T. 1966. Inheritance of drinking behavior. Helsinki. *The Finnish Foundation for Alcohol Studies.* Vol. 14.
10. SPICKETT, S. G. and THODAY, J. M. 1966. Regular responses to selection, III, Interaction between located polygenes. *Genet. Res.* **7,** 96.
11. VANDENBERG, S. G. 1962. The hereditary abilities study : hereditary components in a psychological test battery. *Am. J. hum. Genet.* **14,** 220.
12. VANDENBERG, S. G. 1965. Innate abilities : one or many ? A new method and some results. *Acta Genet. med. Gemell.* **14,** 41.
13. VANDENBERG, S. G. 1968. In defense of twins and the twin method. *Acta psychiat. scand.* (In press.)

GENETIC INFLUENCES

Chairman: DR. C. O. CARTER

GENETICS AND PERSONALITY

H. J. Eysenck

Institute of Psychiatry, University of London

ALTHOUGH some fundamentalists still oppose the notion that individual differences in intelligence are largely determined by genetic causes, most informed judges have given up the Watsonian position of complete environmental determination of such differences as untenable.[46] Such is not the case in relation to non-cognitive aspects of personality, in spite of occasional optimistic statements, such as the following from J. Hirsch:[25] " I shall assume that the battle to overcome ignorance and the behaviouristic opposition to according heredity its proper place in the behavioural sciences has been won effectively and decisively " (p. 119). As a counter to such views we may perhaps quote Redlich and Freedman's[34] recently published and very influential textbook of psychiatry: " The importance of inherited characteristics in neuroses and sociopathies is no longer asserted except by Hans J. Eysenck and D. B. Prell "[18] (p. 176) —the reference being to an experimental study by these two authors which we shall encounter again later on. This assertion indicates well the curious climate of opinion in this field, a climate of opinion more favourable to the " premature crystallization of spurious orthodoxies " than to careful factual study, unimpassioned scrutiny of complex issues and the setting down of tentative conclusions.

What are the reasons for opposition to recognition of the fast-growing body of evidence in favour of the importance of genetic causes in personality development? Hirsch[25] blames J. B. Watson and the behaviourist school; he demonstrates convincingly that Watson committed egregious errors in his discussion of the genetic evidence. But Watson's errors do not implicate behaviourism; and, while some behaviourists have followed Watson in this belief, others have not; behaviourism is a term denoting a methodology, not a body of belief. More responsible, perhaps, is Freudianism, with its implicit—and

163

often explicit—claim to prevent or cure all the behavioural ills our flesh is heir to. If such claims were true, environment would indeed be all powerful; but both Watson and Freud have failed to live up to their promises, as we all know now. Perhaps the real culprit is the optimistic, manipulative climate of opinion, so characteristic of large, growing, technological cultures, which the USA shares with the USSR, the two bastions of anti-hereditarian demagogy. In any case, while there has been a rapid growth of interest in this field, and a rather less rapid growth of worthwhile research, the standard of reviewing in psychiatric and psychological books dealing with personality has not kept pace with these developments. Authors have clung stubbornly to outmoded conclusions and methodological considerations which might have made sense fifty years ago but are not in line with modern trends in genetics.

Much is still made of the conclusions reached by Newman, Freeman and Holzinger (1937) in their classic study of the physical intellectual, educational and personality development of twins. They concluded that " the physical characteristics are least affected by the environment; that intelligence is affected more; educational achievement still more; and personality or temperament, if our tests can be relied upon, the most. " The qualifying phrase is well taken; the personality tests used by these authors were neither reliable nor valid, and no firm interpretation could be based on them. Furthermore, the Woodworth-Mathews Personality Inventory, on which they place the most reliance, was, like their other tests, a measure of temperament for use with adults; Freeman used it on subjects whose mean age was thirteen, and some of whom appear to have been as young as eight! Our own efforts to construct personality tests for children have shown quite clearly that many of the terms used in this and some of the other tests used by Freeman are unfamiliar to children even of twelve or thirteen.[20] Even so, the results show that with the Woodworth-Mathews Inventory, which is a measure of neuroticism, the intraclass correlation is 0·562 for identical twins, 0·371 for fraternal twins, and 0·583 for identical twins brought up in separation. Note particularly the fact that identical twins brought up in separation are slightly more alike than identical twins brought up together; more recent work has shown similar results. Note also the lower correlation for the fraternal twins, which would seem to indicate some definite contribution by heredity to personality differences. The authors comment, unaccountably, that this test "appears to show no very definite trend in correlations, possibly because of the nature of the trait and also because of the unreliability of the measure ". This conclusion, which goes counter to the evidence it is meant to summarize, has often been quoted by later writers, without reference to the actual facts of the situation; in this way do myths in science originate!

In surveying the many studies in this field which have been reported since the Newman, Freeman and Holzinger one, we can follow one of two methods. It would be possible to note briefly each of the many different questionnaires and other tests which have been used with identical and fraternal twins, and comment on the degree of heritability disclosed. As many of the instruments frequently used (such as the MMPI, for instance, or the Gough California Psychological Inventory) contain dozens of sub-scales, such a task would be incapable of being carried out within the time limits of this paper, and even if it could be done, it would prove of doubtful value to the reader in view of the lack of proven validity for the many sub-scales involved. The alternative would be to attempt to structure the field around a limited number of specific hypotheses in the personality field, and examine the literature with respect to hypotheses. It is this latter approach which I have here attempted; the former approach has been followed by Fuller and Thompson,[21] whose much longer report may be consulted with confidence. There are in principle certain advantages which attach to this approach, and I would like to suggest that to some degree it does away with the unscientific shot-gun, haphazard, ad hoc type of research which has made interpretation of findings in this field so difficult. Some of these advantages will become apparent as we proceed.

Recent years have seen a convergence along many different lines on a model of personality which emphasizes two main dimensions, independent of each other, and along each of which a person may occupy any position from one extreme to the other, but with a higher probability attaching to a middle position. These two dimensions are variously called, the one 'neuroticism', 'emotionality', or 'instability', the other 'extraversion-introversion'. Figure 1 illustrates the

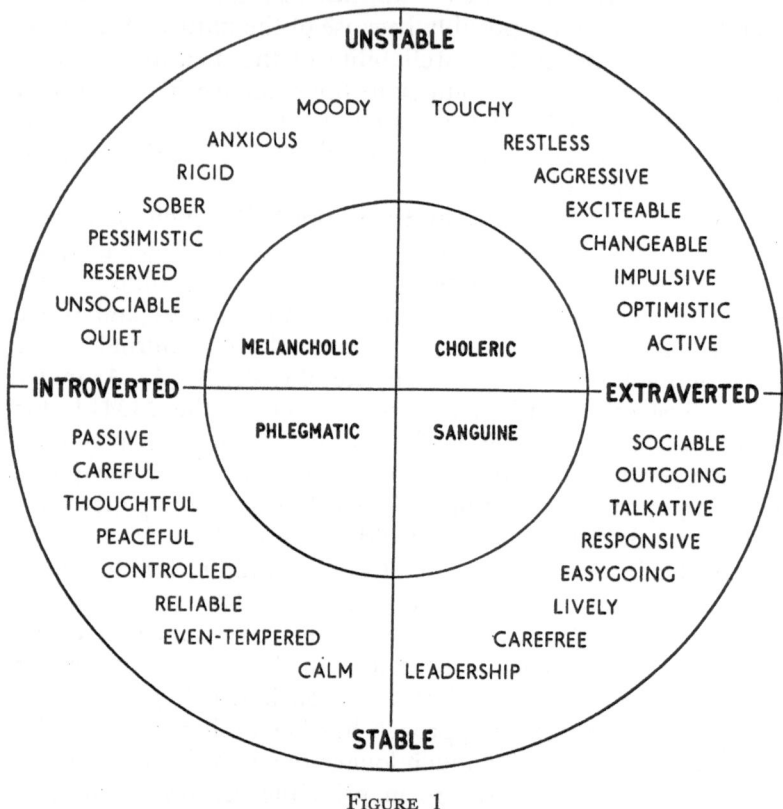

FIGURE 1

TWO MAJOR DIMENSIONS OF PERSONALITY (INTROVERSION-EXTRAVERSION AND EMOTIONAL INSTABILITY-STABILITY), TOGETHER WITH TRAITS INDICATIVE OF THESE MAJOR DIMENSIONS

position, and sets around the rim of the circle some of the traits which characterize these dimensions. Inside the quadrants have been appended the names of the ancient 'four temperaments' which correspond to these particular combinations of the two dimensions: i.e. unstable extraverts correspond to Galen's ' cholerics ', unstable introverts to his ' melancholics ', and so forth. Eysenck and Eysenck[19] have recently assembled the evidence to show that even the super-ficially divergent systems of Cattell and Guilford converge towards a solution formally identical with the one outlined above; and many other writers have arrived at similar con-clusions, although often using a different kind of nomenclature.[14]

It is interesting to note that this scheme coincides in a certain manner with important social classifications which society im-poses on various groups of persons. Thus, neurotic patients receiving treatment in mental hospitals or from private psychia-trists are largely found in the melancholic quadrant, while delinquents and criminals are equally frequently found in the choleric quadrant.[15] It might be thought that perhaps the personality classification was a result, rather than a cause, of the neurotic or delinquent behaviour observed, but this is not so; Burt,[3] for instance, has shown that when 763 children were rated on these two personality dimensions by their teachers and followed up over some thirty-five years, 15 per cent and 18 per cent respectively became habitual offenders or neurotics. Of those who became habitual offenders, 63 per cent had been rated as high on emotionality, and 54 per cent as high on extraversion; only 3 per cent were rated high on introversion. Of those who became neurotics, 59 per cent rated high on emotionality, and 44 per cent as high on introversion; only 1 per cent had been rated high on extraversion. Thus even the probably rather unreliable ratings, made early in life by teachers, predict with surprising accuracy the future delinquent behaviour of the choleric, or the future neurotic breakdown of the melancholic.

We may now look at the experimental literature on the putative hereditary determination of these personality dimen-sions. We may with advantage group the evidence according to certain methodological principles, and start with studies of normal twins, investigated by means of personality question-naires. We have already noted the Newman, Freeman and Holzinger study, and the fact that in spite of their own summing up it did in fact show substantial differences between identical and fraternal twins on the trait of neuroticism. Carter[5] found

M

similar results with the Bernreuter Personality Inventory on a group including high-school, as well as more mature, subjects; his results suggest partial hereditary determination for extraversion as well as for neuroticism. Gottesman[22, 23] used the MMPI and the Cattell scales in two studies also supporting this conclusion. In another study[23] he used the Gough California Psychological Inventory on schoolchildren, discovering strong evidence for heritability of two traits called extraversion-introversion and dependability-independability, a trait resembling neuroticism. So did Wilde[49] in Holland, using a Dutch version of the Maudsley Personality Inventory. Partanen, Brunn and Markkanen[32] in Finland studied 902 male twins aged twenty-eight to thirty-seven, comprising a complete selection of all twins in these age groups; they discovered a heritability index of 0·47 for a sociability trait closely resembling extraversion. Scarr[36] obtained supporting evidence with the Gough Adjective Check List and the Fels Child Behaviour Scales, as did Vandenberg, Stafford et al.[47] on extraversion using the Myers-Briggs Type Indicator.

The most important study in this category, however, is undoubtedly that of Shields,[41] who studied intelligence, extraversion and neuroticism in fraternal twins, and in two groups of identical twins, brought up together or in separation from each other, respectively. His results showed, in brief, that with respect to both cognitive and non-cognitive personality dimensions identical twins were much more alike than fraternal twins, and identical twins who had been brought up separately somewhat more alike than identical twins who had been brought up together. His results are thus similar to those of Newman, Freeman and Holzinger, but more clear-cut and convincing because of his superior psychological approach.

Our second category consists of studies of normal twins using experimental laboratory tests of personality, combined according to the results of factorial analysis of their intercorrelations. Eysenck and Prell,[18] in the study referred to by Redlich and Freedman,[34] found an unusually high heritability coefficient of 0·81 for the neuroticism factor score, and Eysenck[13] found a slightly lower one for extraversion in another, similar study. These values are higher than those found in the questionnaire studies; the reason may be (1) that laboratory tests are closer to the genotype than are questionnaire responses, and (2) that factor analysis succeeds in bringing together the scattered parts of the ' true ' variance more efficiently than the often somewhat arbitrary combination of scores used by questionnaire constructors. Further work along these lines would seem to be urgently needed to verify or disprove these hypotheses, and to replicate the findings, if possible.

A third approach makes use of concordance rates of known criminals or neurotics. Do identical twins of known criminals or neurotics show greater incidence of crime or neurosis than do fraternal twins? Eysenck[15] has summarized a number of studies using some 500 twin pairs; while the results are not quite as striking as the early work of Lange[29] suggested, yet they demonstrate quite convincingly that identical twins have a concordance rate about twice as high as do fraternal twins. Shields[40] found similar results for neurotic maladjustment in youths between twelve and fifteen years of age, with concordance rates over twice as high for identical twins. Shields and Slater[42] have extended this approach to adult neurotics; their results were analyzed according to the specificity of diagnosis. For the most general type of diagnosis (" any coded psychiatric abnormality in co-twin ") the MZ/DZ ratio was 1·7, but for the most specific ("both twins same diagnostic code; smallest subdivision used ") the ratio was 9·3! In other words, not only is there strong evidence of hereditary determination of mental illness, but this hereditary determination seems to extend even to the specific detail of the type of mental illness involved.

All these methods make use of twins, and some objections have been raised to twin research; these will he discussed later on. A fourth method, not based on the differences between identical and fraternal twins, makes use of differential familial resemblances.[4, 6, 7] Lienert and Reisse[30] in Germany found neuroticism correlations between parents to be 0·17; between father and child, 0·13; and between mother and child, 0·31. Calculations from thesefi gures suggested that heredity contributed something like 50 per cent to individual differences, a figure which, however, depends on certain assumptions regarding degree of dominance. Other authors finding not dissimilar results are Hoffeditz,[26] Crook and Thomas,[10] Sward and Friedman[44] and Crook.[9] Results from these studies support the conclusions derived from twin investigations.

A fifth method was used by Brown[2] who studied the incidence of neurosis among parents and sibs of three groups of neurotic patients diagnosed as anxiety neurotics, hysterics and obsessionals. His results showed that close relatives of neurotics are themselves neurotic more frequently than chance would allow, and it also appears that the form of neurosis manifested is rather closely related to that shown by the proband, indicating a certain degree of genetic specificity. A more recent study by Coppen, Cowie and Slater[8] gave rather different results when a personality questionnaire was used; Eysenck[16] has attempted to suggest reasons for their failure to support an hypothesis so widely verified by other authors.

Several other methods of study have been used in the endeavour to clarify the role of a constitutional factor in personality. Thus, physique has been shown to be determined strongly by hereditary causes, and the well-known correlations between extraverted personality and pyknic-athletic body-build[14] may be interpreted as support for a hereditarian view. This relationship is observed quite early in the life of the child,[28,48] suggesting that it is heredity that determines both physique and temperament, rather then physique that determines temperament. Again, the fact that consistency of conduct can be observed at an early age[45] argues for a constitutional type of theory. Breeding studies in animals[1, 17, 39] suggest more specific ways of determining the precise manner of inheritance of behaviour patterns related to the concept of 'personality', as well as affording possibilities of relating these to physiological, neurological and biochemical variables. However, when all is said and done, the main burden of direct proof rests undoubtedly on the twin studies cited, and consequently we must consider the objections which have been raised against them.*

* Inbreeding studies, such as that reported by Schull and Neel[38] with respect to intelligence, would undoubtedly produce valuable additional evidence; unfortunately no such studies have come to hand in the field of non-cognitive personality traits. It should not be difficult to remedy this state of affairs, as personality questionnaires are even easier to administer than are intelligence tests.

It is sometimes objected that diagnosis of zygosity may give rise to error, thus invalidating twin research. There are three answers. In the first place, such errors, if they occurred, would lower the index of heritability; thus the more error, the lower the (apparent) contribution of heredity. In the second place, modern methods of blood-typing, finger-print analysis and so forth are very near perfectly valid. In the third place, even the most impressionistic methods, based on simple inspection of physique, hair colour and other obvious similarities and dissimilarities, are surprisingly accurate when checked against proper tests; Nichols and Bilbro (unpublished)[31] have shown that a simple questionnaire regarding twin similarity had a 93 per cent accuracy when checked against blood-typing. This objection has little value.

More to the point is the fact, demonstrated by Jones,[27] Smith,[43] and Scarr,[35] that identical twins are more similar in the treatment they recieve from their parents than are fraternal twins. The facts are not in dispute, but the interpretation is. Scarr[37] has discussed the two possibilities.

If parents are simply reacting to the existing differences between their DZ twins' behavior, then no bias is introduced into twin studies. But, if they effectively train twin differences, then these environmentally determined differences would bias the comparisons of intraclass correlations in favor of genetic hypotheses, by reducing the possible similarities of DZ co-twins. By the same token, the parents of MZ twins who know their twins are identical may react to existing similarities or seek to train greater similarities than would otherwise exist. When parents are correct about their twins' zygosity, it is impossible to distinguish between parental behavior that is a reaction to the phenotypic behavior of their twins and parental treatment that seeks to train greater differences or similarities.

Fortunately not all parents are correct about their twins' zygosity, and

by examining the cases of parents who are *wrong* about their twins' zygosity, it is possible to separate parental reactions to similarities and differences based on *genetic relatedness* from parental behaviors which arise from their *belief* that their twins should or should not be similar.

Using various rating scales, check lists and other devices, she came to the conclusion that

differences in the parental treatment that twins receive are much more a function of the degree of their genetic relatedness than of parental beliefs about ' identicalness ' and ' fraternal-ness '. . . . The comparisons of parental behavior for correctly and incorrectly classified pairs suggest that environmental determinants of similarities and differences between MZ and DZ co-twins are not as potent as the critics charge.

Along similar lines, it is sometimes alleged that identical twins are closer to each other socially than are fraternal twins, and that they are consequently exposed to similar environments. Shields[40] found that MZ twins were indeed closer to each other socially than were DZ twins, but this degree of closeness was not related to degree of concordance. The evidence, therefore, does not suggest that this criticism is of great relevance to the genetic hypothesis.

The statistics used to estimate heritability are often criticized, particularly Holzinger's ratio, and it seems agreed that it over-estimates the contribution of heredity. In addition, of course, it deals only with the proportion of variance produced by genetic differences *within families*. It is not likely that this restriction imposes too serious a limitation on analysis. Serious efforts are going on at the moment to improve the statistical handling of twin and familial data, and better methods should be available quite soon.

Another criticism suggests that twins are different from the rest of the population from which they are taken. Twins are certainly slightly less intelligent than are single children; with respect to non-cognitive personality variables little is known, but it seems doubtful if this point can be considered very relevant even if true. Slight differences in mean should not upset an analysis of variances and covariances, and it seems highly unlikely that differences between twins and single persons could be anything but slight, even if they exist.

More serious is the suggestion of lack of comparability of intra-uterine environment, with the implication that MZ twins may have more similar environment. The facts do not support such a view; Price[33] has put forward a convincing case in favour of the view that the prenatal environment of fraternal twins is probably more nearly alike than it is for identical twins. If this is true, then we would tend consistently to *under*estimate the importance of heredity in our twin studies.

A similar conclusion is suggested by a consideration of the fundamental notion that MZ twins share identical heredity, a notion which underlies much of the argumentation in this field, whether qualitative or quantitative. As Darlington[11] points out, " this assumption can be justified by authority, by axiomatic authority; but it cannot be justified by experimental evidence assisted by the usual process of inference. Various situations are known in which it is groundless and indeed positively false. The two one-egg twins must differ on genetic grounds; that is to say they must differ internally at the beginning. " Such cases may be arranged fairly neatly in three classes as follows: *a*, nuclear differences; *b*, nucleocytoplasmic differences; and *c*, cytoplasmic differences. Darlington[12] concludes his discussion of these differences by saying: " We have to admit that comparisons of one-egg and two-egg twins do not give us the uncontaminated separation of heredity and environment which Galton and his successors have hoped for. The measure they give is an estimate: but it must almost be an *under*estimate of the effects of heredity."

We may conclude that the errors and criticisms involved in twin studies may partly balance out, but that there appears a likelihood that the estimates of heritability arrived at by the authors cited are underestimates rather than overestimates of the influence of heredity. Many critics doubt the value of such estimates, on the grounds that much uncertainty attaches to them. This point of view seems to go counter to the usual mode of progress in science, where ' ranging shots ' of no great accuracy are followed by careful, painstaking efforts to make estimates and measurements more and more accurate. Thus, for instance, J. D. Van der Waals and Maxwell, by examining the deviations from the ideal-gas laws, were able to determine the number of atoms per unit volume in a gas (Avogadro's number) to an accuracy of about 50 per cent thus making possible eventually an estimation of the size of atoms. Provided we are aware of the likely sources and limits of the errors involved, our present 'ranging shots' would seem to serve a very useful purpose, and to be almost certainly more accurate than the Van der Waals-Maxwell estimate!

One additional reason why many people distrust the evidence in favour of hereditary causes of individual differences

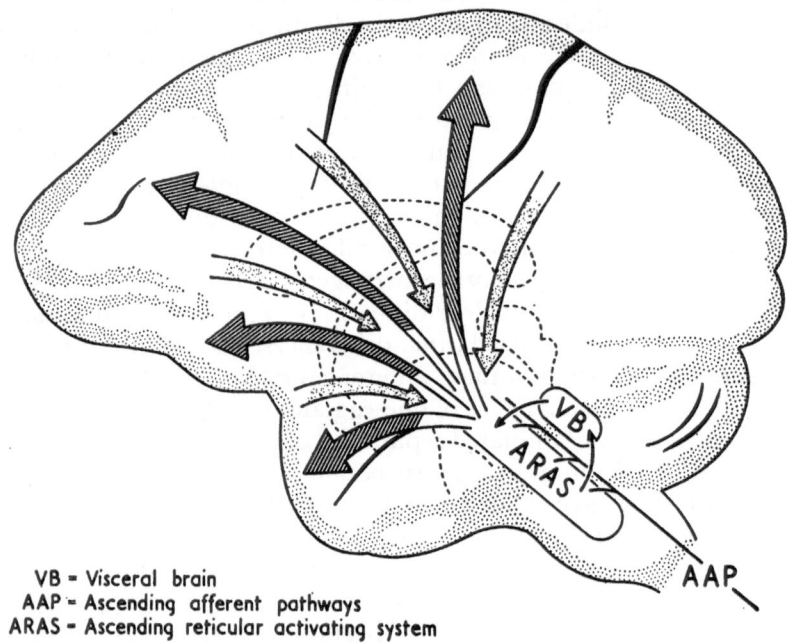

VB = Visceral brain
AAP = Ascending afferent pathways
ARAS = Ascending reticular activating system

FIGURE 2

DIGRAMMATIC REPRESENTATION OF PHYSIOLOGICAL STRUCTURES UNDERLYING
TWO MAJOR DIMENSIONS OF PERSONALITY

in personality is the undoubted fact that we cannot inherit function, but only structure; anatomical and physiological-neurological systems may have their nature determined by the genes, but it is difficult to see how behaviour could be so determined in any direct fashion. The failure to link up behaviour and its putative biological determinants has made hereditary theories regarding the former suspect and somewhat insubstantial. It has recently become possible to remedy this defect by linking both neuroticism and extraversion-introversion with certain anatomical structures of well-known physiological import. As shown in Figure 2 this theory links the emotional over-reactivity so characteristic of neuroticism with constitutionally determined lower thresholds of responsiveness in the visceral brain; this part of the limbic system, as is well known, co-ordinates and organizes the activities of the autonomic system, which is responsible for the expression of our emotions. As regards extraversion-introversion, this has been linked conceptually and experimentally with the ascending reticular arousal system, which is responsible for maintaining the cortex in a state of arousal, thus enabling it to function properly and to deal with incoming sensory messages. The theory maintains that introverts are characterized by a high state of cortical arousal, mediated by low thresholds for incoming sensory stimulation, while extraverts are characterized by a low state of cortical arousal, mediated by high thresholds for incoming sensory stimulation.[16] There is much direct evidence for these two generalizations, both from physiological sources (EEG recordings, evoked potentials, etc.) and from psychological experiments on learning and conditioning, perception, sensory thresholds, motor movements and much else. There is no time to go into the details of this particular theory, but if it should be found to be in any sense along the right lines, then we would here have the much needed link between behaviour and personality, on the one hand, and heritable anatomical structures, on the other. These structures influence behaviour through their interaction with the environment, thus providing us with the kind of model shown in Figure 3. It will be clear from this why experimental laboratory determinations of personality tend to show greater heritability than do questionnaire and rating studies; the former are closer to the hypothetical arousal function (here labelled ' excitation-inhibition balance ') than the latter, which are much more contaminated by environmental influences.

If the evidence for the inheritance of individual differences in personality is so strong, at least in relation to neuroticism-stability and extraversion-introversion, why has it not been universally recognized, and why can psychiatric textbooks (particularly American ones) still be written with an implicit or explicit denial of the importance of any sources of behaviour differences other than environmental ones? The answer was given long ago by the famous physicist Max Planck, who wrote in his *Autobiography*: " A new scientific truth does not triumph by convincing its opponents and making them see the light, but

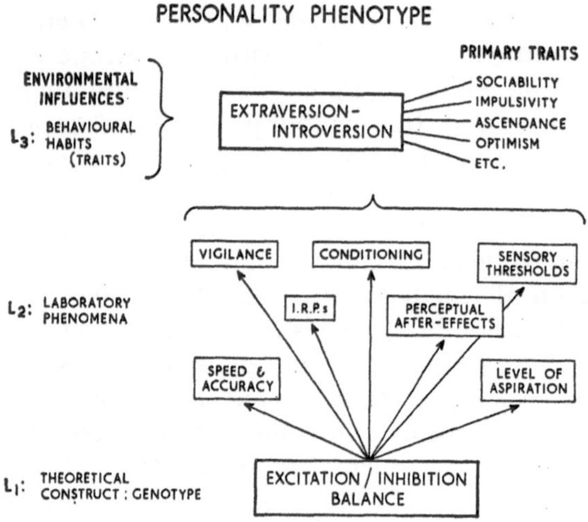

FIGURE 3

DIAGRAM SUGGESTING RELATIONSHIP BETWEEN PER-
SONALITY GENOTYPE AND PHENOTYPE IN RELATION
TO EXTRAVERSION-INTROVERSION

rather because its opponents eventually die, and a new genera-tion grows up that is familiar with it. " If this is so in physics, how much more true it is in psychology, where even the very notion that assertions should have some empirical backing is only slowly gaining ground, and where doctrinaire excesses are still the small coin of exchange of opinion between different ' schools '.

We may thus conclude that the evidence is almost un-animous in stressing the importance of heredity in determining individual differences in personality, without wishing to deny in any way the complementary importance of environment, particularly when environmental circumstances are excessively deviant from the average obtaining in a particular community. The old battle-lines between 'hereditarians' and 'environmentalists' have never had much reality; there apparently are still many of the latter who deny heredity any role in relation to personality, but it would be difficult indeed to find any 'hereditarians' who deny the importance of environment. All we assert, and all that the evidence entitles us to say, is well expressed in the following lines from Thomas Hardy's poem *Heredity*:

> I am the family face;
> Flesh perishes, I live on,
> Projecting trait and trace
> Through time to finis anon,
> And leaping from place to place
> Over oblivion.

REFERENCES

1. BROADHURST, P. L. 1960. Application of biometrical genetics to the inheritance of behaviour. In *Experiments in Personality*. Ed. H. J. Eysenck. London. Routledge and Kegan Paul.
2. BROWN, F. W. 1942. Heredity in psychoneuroses. *Proc. R. Soc. Med.* **35**, 785.
3. BURT, C. 1965. Factorial studies of personality and their bearing on the work of the teacher. *Br. J. educ. Psychol.* **35**, 368.
4. BURT, C. and HOWARD, M. 1956. The multifactorial theory of inheritance and its application to intelligence. *Br. J. statist. Psychol.* **9**, 95.
5. CARTER, H. P. 1933. Twin similarities in personality traits. *J. genet. Psychol.* **43**, 312.
6. CATTELL, R. B. 1953. Research designs in psychological genetics with special reference to the multiple variance method. *Am. J. hum. Genet.* **5**, 76.
7. CATTELL, R. B. 1963. The interaction of hereditary and environmental influences. *Br. J. statist. Psychol.* **16**, 191.
8. COPPEN, A., COWIE, V. and SLATER, E. 1965. Familial aspects of 'neuroticism' and 'extraversion'. *Br. J. Psychiat.* **111**, 70.
9. CROOK, M. N. 1937. Intra-family relationship in personality test performance. *Psychol. Rec.* **1**, 479.
10. CROOK, M. N. and THOMAS, M. 1939. Family relationships in ascendance-submission. *Univ. Calif. Publs. Educ. Philos. Psychol.* **1**, 189.

11. DARLINGTON, C. D. 1954. Heredity and environment. *Caryologia* 6 (Supplem.) (*Proc. IX Int. Conf. Genet.*), p. 370.
12. DARLINGTON, C. D. 1963. Psychology, genetics and the process of history. *Br. J. Psychol.* 54, 293.
13. EYSENCK, H. J. 1956. The inheritance of extraversion-introversion. *Acta psychol.* 12, 95.
14. EYSENCK, H. J. 1960. *The Structure of Human Personality.* London. Methuen.
15. EYSENCK, H. J. 1964. *Crime and Personality.* Boston. Houghton Mifflin.
16. EYSENCK, H. J. 1967. *The Biological Basis of Personality.* Springfield. C. C. Thomas.
17. EYSENCK, H. J. and BROADHURST, P. L. 1964. Experiments with animals. In *Experiments in motivation.* Ed. H. J. Eysenck. New York. Pergamon.
18. EYSENCK, H. J. and PRELL, D. 1951. The inheritance of neuroticism: an experimental study. *J. ment. Sci.* 97, 441.
19. EYSENCK, H. J. and EYSENCK, S. B. G. 1968. *The Description and Measurement of Personality.* London. Routledge and Kegan Paul.
20. EYSENCK, S. B. G. 1965. *Manual of the Junior Eysenck Personality Inventory.* London. Univ. London Press; San Diego. Educ. Indust. Testg. Service.
21. FULLER, J. L. and THOMPSON, W. R. 1960. *Behaviour Genetics.* London. Wiley.
22. GOTTESMAN, I. I. 1963. Heritability of personality: a demonstration. *Psychol. Monogr.* 77, 9.
23. GOTTESMAN, I. I. 1965. Genetic variance in adaptive personality traits. Paper presented to Amer. Psychol. Ass.
24. GOTTESMAN, I. I. 1965. Personality and natural selection. In *Methods and Goals in Human Behaviour Genetics* (pp. 63–80). Ed. S. G. Vandenberg. New York. Academic Press.
25. HIRSCH, J. 1967. Behavior—genetic, or ' experimental ' analysis. *Am. Psychol.* 22, 118.
26. HOFFEDITZ, E. L. 1934. Family resemblances in personality traits. *J. soc. Psychol.* 5, 214.
27. JONES, H. E. 1955. Perceived differences among twins. *Eugen. Q.* 5, 98.
28. KAGAN, J. 1966. Body build and conceptual impulsivity in children. *J. Personality* 34, 110.
29. LANGE, J. 1931. *Crime and Destiny.* London. Allen and Unwin.
30. LIENERT, G. A. and REISSE, N. 1961. Ein korrelations-analytischer Beiträg zur genetischen Determination des Neurotizismus. *Psychol. Beitr.* 7, 121.
31. NICHOLS, R. C. and BILBRO, W. C. The diagnosis of twin zygocity. Unpublished report. National Merit Scholarship Corporation.
32. PARTANEN, J., BRUNN, K. and MARKKANEN, T. 1966. Inheritance of drinking behaviour. *Finnish Foundation for Alcohol Studies.* Vol. 14.

33. PRICE, B. 1950. Primary biases in twin studies. *Am. J. hum. Genetic.* **2,** 293.
34. REDLICH, F. C. and FREEDMAN, D. G. 1966. *The Theory and Practice of Psychiatry.* New York. Basic Books.
35. SCARR, S. 1964. Genetics and human motivation. Unpublished Ph.D. thesis, Harvard Univ.
36. SCARR, S. 1965. The inheritance of sociability. Paper read at Meeting of Am. Psychol. Ass.
37. SCARR, S. 1966. Environmental bias in twin studies. Unpublished Rep., Inst. Child Study, Univ. Maryland.
38. SCHULL, W. J. and NEEL, J. V. 1965. *The Effects of Inbreeding on Japanese Children.* New York. Harper and Row.
39. SCOTT, J. C. and FULLER, J. L. 1965. *Genetics and the Social Behavior of the Dog.* Chicago. Univ. Press.
40. SHIELDS, J. 1954. Personality differences and neurotic traits in normal twin school children. *Eugen. Rev.* **45,** 213.
41. SHIELDS, J. 1962. *Monozygotic Twins.* Oxford. Univ. Press.
42. SHIELDS, J. and SLATER, E. 1966. La similarité du diagnostic chez les jumeaux. *L'Evolution psychiat.* **2,** 441.
43. SMITH, R. T. 1965. A comparison of socio-environmental factors in monozygotic and dizygotic wins. In *Methods and Goals in Human Behavior Genetics.* Ed. S. G. Vandenberg. New York. Academic Press.
44. SWARD, K. and FRIEDMAN, M. B. 1935. Jewish temperament. *J. appl. Psychol.* **19,** 70.
45. THOMAS, A., CHESS, S., BIRCH, H. G., HERTZIG, M. G. and KORN, S. 1964. *Behavioural Individuality in Early Childhood.* London. Univ. London Press.
46. VANDENBERG, S. G. 1966. The nature and nurture of intelligence. *Res. Rep. No. 20, Louisville Twin Study, Child Developm. Unit.*
47. VANDENBERG, S. G., STAFFORD, R. E., BROWN, A. and GRESHAM, J. 1966. *Prog. Rep. No. 15, Louisville Twin Study, Child Developm. Unit.*
48. WALKER, R. N. 1962. Body build and behavior in young children. *Monogr. soc. Res. Child Dev.* **27,** No. 84.
49. WILDE, G. J. S. 1964. Inheritance of personality traits. *Acta psychol.* **22,** 37.

SEX CHROMOSOME ANEUPLOIDY AND CRIMINAL BEHAVIOUR

W. M. COURT BROWN, PATRICIA A. JACOBS and W. H. PRICE

MRC Clinical and Population Cytogenetics Research Unit,
Western General Hospital, Edinburgh

DURING the last eight years knowledge has been acquired of the frequency of persons with an abnormal sex chromosome complement. Much of this has come from the use of nuclear sexing combined with an examination of the chromosomes of those whose nuclear sex is at variance with their phenotypic sex. More recently it has become possible to undertake chromosome studies on a fairly wide scale, and to use these findings as the primary means of ascertaining abnormal individuals. Such studies have the advantage of providing more information than those restricted to nucleur sexing, particularly on the frequency of males with a single X chromosome but an abnormal number of Y chromosomes. This communication will describe in summary form the present state of knowledge of the frequency of sex chromosomally abnormal individuals among those with behavioural disturbances, and it will be confined to the study of phenotypic males, because criminal behaviour among females does not pose so serious a problem.

To understand the significance, or lack of significance, of the findings in such groups as men in maximum security hospitals or young men at Borstal institutions, it is necessary first to discuss the information available on the frequency of males with an abnormal chromosome complement in the ordinary population, and among patients in the ordinary hospitals for the mentally subnormal. After this it will be possible to make some assessment of the results of examining patients in maximum security hospitals and males in corrective training establishments both in this country and in the United States.

We know very much more about the frequency of males in the liveborn population who have an abnormal sex chromosome complement than in any other sample of the general male population.[11] To date there are data from the nuclear sexing of over 41,000 randomly selected male babies from ten surveys carried out in seven different countries (Table I). The results are all reasonably compatible and the frequency of liveborn

TABLE I

THE FREQUENCY OF CHROMATIN-POSITIVE MALES IN
THE LIVEBORN POPULATION

TOTAL MALES	No.* CHROMATIN-POSITIVE	CHROMATIN-POSITIVES PER 1000 LIVEBORN
41,688	71	1·7

Data from 10 surveys from 13 localities in 7 different countries—Canada, India, Poland, Scotland, Switzerland, USA and USSR. (For details see Court Brown.[11])

* All were singly chromatin-positive.

males with an abnormal nuclear sex is about 1·7 per 1000, somewhat higher than that of children with mongolism. As the identification has been by nuclear sexing, these abnormal males are those with more than one X chromosome. So far, all have been singly chromatin-positive, and no instance has yet been recorded of a male baby identified in a survey whose nuclear sex would lead one to suspect the presence of more than two X chromosomes.

Nearly half these babies have been examined in two surveys, those under way in Edinburgh and in Denver; and these two surveys contribute nearly all the information on the chromosome constitutions of the chromatin-positive children, who at birth betray no evidence of their abnormality in their physical development. The XXY is the most frequent abnormal complement, the frequency of XXY babies being about 1·3 per 1000 liveborn. Nearly all the other abnormal children have been mosaics of the XY/XXY type with a frequency of about 0·4 per 1000, so that the ratio of XXY babies to those with an XY/XXY mosaic constitution is about 3 : 1. All other abnormalities are very rare at birth, and this is true of the XXYY complement, which will be referred to again later.

There have been a number of studies on older males in the general population, particularly on hospital patients and military recruits (Table II). These studies show a frequency of singly chromatin-positive males marginally, but not significantly, in excess of the babies. It would be a mistake, however,

TABLE II

NUCLEAR SEXING SURVEYS OF VARIOUS GROUPS OF
NON-INSTITUTIONALIZED MALES

AUTHOR	GROUP EXAMINED	TOTAL MALES	No.* CHROMATIN- POSITIVE	CHROMATIN- POSITIVES PER 1000
Baikie et al.[1]	Hospital admissions, all ages, Melbourne, Australia.	4137	7	7·1
Hambert [13]	Swedish military con- scripts.	2752	6	2·2
Kaplan and Norfleet [16]	US army recruits.	1000	2	2·0
Paulsen et al.[20]	Out-patients, all ages, US public health ser- vice hospital.	977	2	2·0
Baikie et al.[1]	Rural Australian com- munity.	942	6	6·4
—	All groups	9808	23	2·3

* All were singly chromatin-positive.

to accept these data as indicative of the real frequency of chromatin-positive males in the population, for there are indications that the abnormal male babies may have heightened mortality risks,[6, 17] and we know that some chromatin-positive males do become segregated into the hospitals for the mentally subnormal in an unusual frequency. We would expect, therefore, the true frequency to be less than the 1·7 per 1000 found in the liveborn population. In the absence of any better data this latter population is used for purposes of comparison, but it should be remembered that, where such comparisons indicate a significantly increased frequency in a particular group of males, the extent of the increase may be underestimated.

Information is available from fifteen surveys of males in

TABLE III

NUCLEAR SEXING SURVEYS OF MALES IN HOSPITALS FOR THE
MENTALLY SUBNORMAL

TOTAL MALES	No.* CHROMATIN-POSITIVE	CHROMATIN-POSITIVES PER 1000
13,349	126	9·4

Data from 15 surveys and 8 countries—Canada, England, Ireland, Japan Scotland, South Africa, Sweden and USA. (For details see Court Brown et al.[11])

* 101 singly chromatin-positive, 21 doubly positive and 4 trebly positive.

hospitals for the mentally subnormal,[11] covering over 13,000 patients and from eight different countries (Table III). On average the frequency of those with an abnormal nuclear sex is about 9·4 per 1000, or between five and six times that in liveborn male babies. Three of the surveys provide valuable information on frequency in relation to IQ, all three showing very consistent results (Table IV). Among the most severely retarded—those

TABLE IV

MALE PATIENTS IN HOSPITALS FOR THE MENTALLY SUBNORMAL.
THE FREQUENCY OF CHROMATIN-POSITIVE PATIENTS AND IQ [2, 3, 4, 5, 7, 19]

IQ	TOTAL MALES	No.* CHROMATIN-POSITIVE	No. PER 1000
<20	901	2	2·2
20–49	1767	16	9·0
50 or more	864	13	15·2
All ranges	3532	31	8·8

* 23 singly chromatin-positive, 7 doubly positive and 1 trebly positive.

stated to have an IQ of <20—the frequency is 2·2 per 1000. Among those with an IQ between 20 and 50 it is 9·0 per 1000, and among those with an IQ of 50 or more it is about 15·2 per 1000. So among the worst retardates the frequency of abnormal males is not materially different from that among the liveborn male

N

population, while there is about a ninefold increase among the high grade mentally subnormal, of whom, on average, about one male in sixty-five has an abnormal nuclear sex.

All these data have the great disadvantage that we are largely ignorant of the age-distributions of the patients in the hospitals for the mentally subnormal. It would be very surprising, however, if the abnormal males were randomly distributed by age, and it would be expected that the frequency among adult males would be higher than among children. The reasoning behind this stems from our observations that the majority of the abnormal males are committed to the hospitals from the age of fifteen upwards, many, in fact, getting there via the Courts. So, if we could examine the frequency of abnormal males by age among those with an IQ of 50 or more, it would be surprising if the frequency among adults was not found to be in fact higher than the quoted 15·2 per 1000.

Among the mentally subnormal we find males with more than two X chromosomes, the commonest of these being the XXXY male—with a frequency of about 1·1 per 1000 (Table V). As the frequency of XXY males, not corrected for IQ, is about 5·1 per 1000, the ratio of XXY to XXXY males is about

TABLE V

ESTIMATED FREQUENCIES OF CHROMATIN-POSITIVE MALES BY KARYOTYPE
IN HOSPITALS FOR THE MENTALLY SUBNORMAL.[11]

XXY males	5·1 per 1000
XXXY males	1·1 per 1000
XXXXY males	<1·0 per 1000
XXYY males	<1·0 per 1000
XX males	<1·0 per 1000
XY/XXY males	<1·0 per 1000
Other mosaics	1·2 per 1000

5:1. Very significantly there is a considerable increase in the ratio of XXY males to XY/XXY mosaic males—from 3 : 1, as noted in the babies, to about 8 : 1. In fact there is little evidence that the XY/XXY mosaic incurs any special risk of being mentally subnormal. Finally, the position about XXYY males is that they are more frequent than at birth, but even among the mentally subnormal they appear rare, with a frequency of somewhat less than 1·0 per 1000 without taking IQ into account.

Before leaving these hospital investigations the point is worth making that a comparison of the frequencies of XXY males and XXX females in the hospitals and in the newborn suggests that the quantitative effect of an extra X chromosome on intelligence is about the same in both sexes. The ratio of the frequencies for XXY males is about 3·9 : 1 and for XXX females about 4·6 : 1. These could be even closer if the frequencies in the hospitals for the mentally subnormal were compared with those in the appropriate sample of the general population.

We must now turn to the problem of males with behavioural disturbances, first concentrating on males with an abnormal nuclear sex. Forssman and Hambert[12] from Sweden have paid particular attention to males in Swedish hospitals for those described as both mentally subnormal and hard-to-manage, and their work was published in a summary form in 1963 and more fully by Hambert in 1966.[13] The patients in six hospitals were examined, in five of them males with an IQ of 50 or more, and in one those with a lower IQ. By definition, these hospitals are not strictly comparable with the British maximum security hospitals. In Sweden every effort is made to accommodate the mentally subnormal within the general community, but those whose behaviour prevents this are segrated in these hospitals. Many, of course, have criminal records, but the possession of such is not a primary element in their segregation.

There is a remarkable consistency in the results from the five hospitals for men with an IQ of 50 or more (Table VI), with altogether 19 of 958 men having an abnormal nuclear sex or 19·8 per 1000. This is a higher frequency than among males in hospitals for the mentally subnormal with an IQ of 50 or more, possibly because of differences in age structure, but not significantly higher. Nor is the frequency of males with an IQ of <50 importantly different from that found in the ordinary hospitals for the mentally subnormal. Of the men with an IQ of 50 or more more 16 were singly chromatin-positive and 3 were doubly chromatin-positive, the ratio of 5 : 1 again not being importantly different from that in the hospitals for men in the same range of IQ. Unfortunately the chromosome

TABLE VI

NUCLEAR SEXING SURVEYS IN SWEDISH HOSPITALS FOR MENTALLY
SUBNORMAL HARD-TO-MANAGE MALES [13]

IQ 50 or more

HOSPITAL	TOTAL MALES	No.* CHROMATIN-POSITIVE	No. ABNORMAL PER 1000
Salberga	381	7	18·4
Västra Ny.	182	4	22·0
Källshagen	304	6	19·7
Salbohed	52	1	19·2
Vingåker	39	1	25·6
All	958	19	19·8

IQ <50

Vipeholm	620	3	4·8

* IQ 50 or more : 16 singly chromatin-positive and 3 doubly positive.
IQ <50 : 2 singly positive and 1 doubly positive.

examinations are too incomplete for us to look at karyotype
frequencies, and this study really underlines the importance of
associated behavioural disorders in considering whether the high
grade mentally subnormal should be hospitalized.

This is a suitable point to consider whether there is evidence
that males with an abnormal nuclear sex may show an important
degree of behavioural disorder without being classifiable as
mentally subnormal. There is, in fact, some evidence for this
from a study by Mosier and colleagues in California.[19] These
authors had noted in an examination of a hospital for the
mentally subnormal that there appeared to be a significant
excess of sexual crimes among chromatin-positive males than
chromatin-negative males in the same institution, after stand-
ardizing for IQ.[18] They then examined 600 randomly selected
sexual psychopaths from another institution and found six to
be chromatin-positive, i.e. 10 per 1000. The distribution of
the 600 men by IQ is not recorded, but we know something of
the IQs of the abnormal men. Only one was on the borderline
of subnormality with an IQ of 83, the other five being either
measured as normal or assessed as normal (Table VII). Even

TABLE VII

CHROMATIN-POSITIVE MALES AMONG SEXUAL PSYCHOPATHS [13]

PATIENT	IQ	PSYCHIATRIC DIAGNOSIS	OFFENCE
A	83	Personality pattern disturbance : schizoid personality.	Attempted rape
B	101	Personality pattern disturbance : schizoid personality.	Attempted rape
C	Normal	Personality trait disturbance : passive aggressive personality.	Paedophilia
D	95	Personality trait disturbance : emotionally unstable personality.	Sodomy
E	Dull, normal	Sociopathic personality disturbance : sexual deviation.	Paedophilia
F	Normal	Sociopathic personality disturbance : sexual deviation.	Paedophilia

if none of the 600 were subnormal, the frequency of chromatin-positive males is suspiciously high, and it would be even more striking given there was a subnormal component, and the abnormal males were related to the normal component. So these findings suggest that behavioural disorders, certainly in respect of sexual misdemeanours, may be dissociated from mental subnormality in chromatin-positive males.

Before dealing with the maximum security hospitals, there are four other surveys of criminal groups to be noted. The first three are nuclear sexing surveys and the fourth a chromosome survey (Table VIII). The first is the one made by de la Chapelle and colleagues[10] in Finland of a group of criminals, no other specification being given, in which one of 383 males was chromatin-positive. The second and third are by Wegmann and Smith[14] from Wisconsin. One was of juvenile delinquents of twelve to eighteen years at the Wisconsin State hospital for boys, with none of 505 having an abnormal nuclear sex. The other was of young adult males at the Wisconsin State Reformatory, with two of 813 being chromatin-positive. The fourth study, an unpublished one by Jacobs and her colleagues in Edinburgh, was of all the males sentenced for Borstal training in Scotland during a twelve-month period in 1966 and 1967: 607 boys had their chromosomes examined and three were chromatin-positive, two XXY males and an XX male. The

TABLE VIII

SURVEYS OF VARIOUS MALE CRIMINAL GROUPS

AUTHOR	TYPE OF SURVEY	GROUP	TOTAL MALES	No.* CHROMATIN-POSITIVE
de la Chapelle[10]	Nuclear sex	Criminals, no other specification.	383	1
Wegmann and Smith[24]	Nuclear sex	Juvenile delinquents: Wisconsin State hospital.	505	0
	Nuclear sex	Young adult males : Wisconsin State reformatory.	813	2
Jacobs et al. (unpublished)	Chromosome	New entrants : Scottish Borstals 1966–1967.	607	3

* All singly chromatin-positive.

findings of all four surveys are compatible and, in all, over 2300 males were examined and 2·8 per 1000 were found to be chromatin-positive. At least we can say that these studies suggest no important increase in abnormal males in these groups, particularly among those attending schools or institutions for corrective training.

The position, however, is quite different in the maximum security hospitals. Putting together Moss Side and Rampton in England, studied by Casey and colleagues,[8] and Carstairs in Scotland, studied by Jacobs and colleagues,[15] 24 of 1257 men were chromatin-positive, or 19·1 per 1000 (Table IX). All were singly chromatin-positive. At this point comparisons become difficult; for, while the majority were classifiable as mentally subnormal, there was a substantial fraction that were not. For what it is worth, it was found that the frequency of chromatin-positive males in these hospitals is greater than in the hospitals for the mentally subnormal when the comparison is restricted to singly chromatin-positive males with an IQ of 50 or more. In the ordinary hospitals for the mentally subnormal the frequency of these is 11·6 per 1000. The increase in frequency in the maximum security, between one and a half and twice what it is in the other hospitals, is not significant.

TABLE IX

CHROMATIN-POSITIVE MALES IN MAXIMUM SECURITY HOSPITALS
AT MOSS SIDE, RAMPTON AND CARSTAIRS [9, 15]

TOTAL MALES	No.* CHROMATIN- POSITIVE	CHROMATIN- POSITIVES PER 1000	CHROMOSOME CONSTITUTIONS
1257	24	19·1	XXY (14), XXYY (8) XY/XXY (1) XY/XXY/XXY (1)

In Carstairs 12 of 315 males had an abnormal chromosome complement, including 9 XYY males, or 38·1 per 1000.

* All singly chromatin-positive.

Perhaps it would be with larger numbers or a more critical comparison in relation to IQ. However, what appeared to us important about the maximum security hospitals was the apparently high proportion of XXYY males. These were a third of the chromatin-positive males although it must be confessed that we had not realized at the time that these males appeared to form as high a proportion of the singly chromatin-positive males with an IQ of 50 or more in the ordinary hospitals for the mentally subnormal. At any rate, the findings in the maximum security hospitals led Jacobs and her colleagues to postulate, in the event correctly but for the wrong reasons, that an extra Y chromosome might influence behaviour and that the XYY male might be unusually frequent among criminal psychopaths. This led to a search for such men among those in the State hospital in Carstairs through a chromosome survey of the patients in that hospital.

The results of this search are now well known and will only be briefly recapitulated.[14, 15] About 3 per cent of men were found to have an XYY complement (Table X), and, in fact, counting all abnormalities of the sex chromosome complement, these had a frequency of 38·1 per 1000. The finding of 3 per cent with an XYY complement must represent a very substantial increase in frequency over these males at birth. The determination of the birth frequency is going to be very tedious as it requires the chromosome study of many thousands of randomly selected male babies. We have made a start in Edinburgh and so far have found one of 618 male babies to have an XYY complement. There are theoretical reasons for expecting that their frequency should be less than that of XXY males, which is 1·3 per 1000.

TABLE X

INFORMATION ON XYY MALES IN MAXIMUM SECURITY HOSPITALS [9, 14, 15]

HOSPITAL	GROUP	TOTAL MALES	No. XYY	XYY's PER 1000
Carstairs	All males *	315	9	28·6
	183 cm or more	21 } 315	5	238·1
	<183 cm	294 }	4	13·6
	Mentally subnormal	208 } 315	7	33·6
	Mentally normal	107 }	2	18·7
Moss Side and Rampton	183 cm or more	50	12	240·0
Broadmoor	183 cm or more	50	4	80·0

* Does not include 27 males who were not prepared to co-operate. The distribution of height of these did not differ from that of those co-operating, but they were significantly more intelligent.

The XYY males were physically normally developed, but we know nothing about their reproductive performance.[21] Their distribution of stature was unusual as, on average, they were found to be about 10 cm. taller than the XY male in the same institution. Jacobs and her colleagues found one in four of the men who were 183 cm. (6 feet) or more tall to have an XYY complement, a finding amply confirmed on larger numbers of men at Moss Side and Rampton by Casey and his colleagues.[9] They were found among those assessed as not mentally subnormal as well as among the mentally subnormal, a finding also confirmed by Casey and colleagues on the men at Broadmoor. The evidence, so far, indicates that it is unlikely that an extra Y chromosome will have a particularly marked effect on IQ, and probably not as pronounced an effect as an extra X chromosome. It is perhaps unlikely that XYY males will be found in any important numbers in hospitals for the mentally subnormal except in hospitals that accept patients under a Court order.

An inquiry into the family histories of the men at Carstairs and of matched controls with an XY complement strongly suggested the XYY males to be genetically predisposed towards behavioural disorders, and that environmental circumstances were not of primary importance.[22, 23] The burden of their crimes was against property rather than against the person, although aggressive behaviour was not unknown among the XYY males at Carstairs. Since this study, we have had the opportunity of studying other cases, and we have also had reports of yet other cases from abroad. It is beginning to look as if the XYY males are subject to a wide variation of psychiatric disturbances. They do not all necessarily commit crimes, nor, indeed, need they all show a disturbance of one kind or another. However, it does look as if an important proportion are disturbed and it seems worth while searching for them among young adolescents or males attending psychiatric clinics.

To sum up, we now recognize that there are three groups of males with an abnormal sex chromosome complement who can show behavioural disorders of such a type as to bring them into conflict with the law. These are XXY males, XYY males and XXYY males. All three groups may show mental subnormality, but for XXY and XYY males there is evidence that in some their mental subnormality can be largely or wholly dissociated from the behavioural disorder. We are not in a position to suggest the relationship between the abnormal chromosome complement, the physical abnormalities and those of behaviour in the XXY male, but there is good evidence to support the idea that the abnormal complement of Y chromosomes may strongly influence the behaviour of the physically normal XYY male. Finally, although sufficient data are not yet to hand, we strongly suspect the existence of differences between the behavioural disorders of XXY and XYY males, and the probability that there is a greater component of aggressive, particularly aggressive sexual, behaviour among those with an extra X chromosome than among those with an extra Y.

REFERENCES

1. BAIKIE, A. G., GARSON, O. M., WESTE, S. M. and FERGUSON, J. M. 1966. Numerical abnormalities of the X chromosome. *Lancet.* i, 398.
2. BARR, M. L., CARR, D. H., MORISHIMA, A. and GRUMBACH, M. M. 1962. An XY/XXXY sex chromosome mosaicism in a mentally defective male patient. *J. ment. Def. Res.* 6, 65.
3. BARR, M. L., CARR, D. H., SOLTAN, H. C., WIENS, R. G. and PLUNKETT, E. R. 1964. The XXYY variant of Klinefelter's syndrome. *Can. med. Ass. J.* 90, 575.
4. BARR, M. L., SHAVER, E. L., CARR, D. H. and PLUNKETT, E. R. 1959. An unusual sex chromatin pattern in three mentally deficient subjects. *J. ment. Def. Res.* 3, 78.
5. BARR, M. L., SHAVER, E. L., CARR, D. H. and PLUNKETT, E. R. 1960. The chromatin-positive Klinefelter's syndrome among patients in mental deficiency hospitals. *J. ment. Def. Res.* 4, 89.
6. BOCHKOV, N. P. 1965. Frequency of sex chromosome abnormalities in newborn and dead children. In *Proc. Symp. on the Mutational Process.* Prague, 1965. Academia, Praha.
7. BREG, W. R., CASTILLA, E. E., MILLER, O. J. and CORNWELL, J. G. 1963. Sex chromatin and chromosome studies in 1562 institutionalized mental defectives, *J. Paediat.* 63, 738.
8. CASEY, M. D., SEGALL, L. J., STREET, D. R. K. and BLANK, C. E. 1966. Sex chromosome abnormalities in two State hospitals for patients requiring special security. *Nature, Lond.,* 209, 641.
9. CASEY, M. D., BLANK, C. E., STREET, D. R. K., SEGALL, L. J., McDOUGALL, D. H., McGRATH, P. J. and SKINNER, J. L. 1966. YY chromosomes and antisocial behaviour. *Lancet,* ii, 859.
10. DE LA CHAPELLE, A. 1963. Sex chromosome abnormalities among the mentally defective in Finland. *J. ment. Def. Res.* 7, 129.
11. COURT BROWN, W. M. 1967. Sex chromosomes aneuploidy in man, with reference to its epidemiology, mental subnormality and behavioural disorders. *Int. Rev. Exp. Path.* (in press).
12. FORSSMAN, H. and HAMBERT, G. 1963. Incidence of Klinefelter's syndrome among mental patients. *Lancet* i, 1327.
13. HAMBERT, G. 1966. *Males with positive sex chromatin.* Akademiforlaget. Goteberg, Sweden.
14. JACOBS, P. A., BRUNTON, M., MELVILLE, M. M., BRITTAIN, R. P. and McCLEMONT, W. F. 1965. Aggressive behaviour, mental subnormality and the XYY male. *Nature, Lond.* 208, 1351.
15. JACOBS, P. A., PRICE, W. H., COURT BROWN, W. M., BRITTAIN, R. P. and WHATMORE, P. B. 1967. Chromosome studies on men in a maximum security hospital. *Ann. hum. Genet.* (in press).
16. KAPLAN, N. M. and NORFLEET, R. G. 1961. Hypogonadism in

young men (with emphasis on Klinefelter's syndrome). *Ann. intern. Med.* 54, 461.

17. MACLEAN, N., HARNDEN, D. G., COURT BROWN, W. M., BOND, J. and MANTLE, D. J. 1964. Sex chromosome abnormalities in newborn babies, *Lancet* i, 286.

18. MOSIER, H. D., SCOTT, L. W. and COTTER, L. H. 1960. The frequency of the positive sex chromatin pattern in males with mental deficiency. *Pediatrics, Springfield* 25, 291.

19. MOSIER, H. D., SCOTT, L. W. and DINGMAN, H. F. 1960. Sexually deviant behaviour in Klinefelter's syndrome. *J. Pediat.* 57, 479.

20. PAULSEN, C. A., DE SOUZA, A., YOSHIZUMI, T. and LEWIS, B. M. 1964. Results of a buccal smear survey in non-institutionalised adult males. *J. clin. Endoc.* 24, 1182.

21. PRICE, W. H., STRONG, J. A., WHATMORE, P. B. and McCLEMONT, W. F. 1966. Criminal patients with XYY sex chromosome complement. *Lancet* i, 565.

22. PRICE, W. H. and WHATMORE, P. B. 1967. Criminal behaviour and the XYY male. *Nature, Lond.* 213, 815.

23. PRICE, W. H. and WHATMORE, P. B. 1967. Behaviour disorders and the pattern of crime among XYY males identified at a maximum security hospital. *Br. med. J.* i, 533.

24. WEGMANN, T. G. and SMITH, D. W. 1963. Incidence of Klinefelter's syndrome among juvenile delinquents and felons. *Lancet* i, 274.

GENES, HORMONES
AND BEHAVIOURAL VARIATION

J. G. M. Shire*

Department of Genetics, University of Cambridge

THE papers by Professor Vandenberg,[27] Professor Eysenck[10], and Dr. Court Brown[7] have all considered genetic influences on behaviour, although the approaches adopted have been different. Genetic differences between normal individuals that affect their behaviour have been discussed, as well as changes in the genetic material which result in pathological disturbances. The most interesting behavioural differences are not those between people suffering from major diseases of the nervous system and normal healthy people, but those differences, such as are found in IQ, between the many individuals which make up every normal population. Many workers have shown that there are important genetic elements underlying these differences.[13]

Genes that bring about differences in behaviour may have their primary actions, at the molecular level, within the central nervous system, thus producing their effects on behaviour directly. Alternatively, their primary sites of action may be in other parts of the body, and their effects on behaviour be brought about through the general physiology of the individual concerned. Hormones are known to have actions on the nervous system and to affect the behaviour patterns of man and other animals.[4, 12] It is also well known that an individual's behaviour can have marked effects on his endocrine system.[6, 15] It is therefore reasonable to expect that some genes which affect hormone systems will also have effects on behaviour. Indeed some major mutants, such as congenital goitre in man[23] and pituitary dwarfism in mice,[11] whose prime action is on the endocrine system, do have marked effects on behaviour. One can therefore ask: is there any evidence of genetic differences

* MRC Research Fellow.

that give rise to variety in the physiology and, particularly, the endocrine systems of normal individuals? While there are some suggestive observations on man,[28] studies on mice are much easier to make, and have the important advantage that they can be followed up by controlled breeding experiments, i.e. genetical ones. Although mice are not identical with men, either in physiology or in behaviour, the general patterns are the same. It is reasonable, therefore, to assume that if there is much genetic variation affecting the endocrine systems of normal mice, there will be analogous variants among normal men.

This paper is concerned with the evidence for the existence of genetic variation affecting the endocrine systems of healthy mice, and with the problems of the relationships between strain differences in behavioural and hormonal characters. The adrenal glands were chosen for study because they produce important hormones like adrenaline and those of the group to which cortisone belongs.

GENETIC VARIATION IN AN ENDOCRINE SYSTEM IN
NORMAL MICE

Observations and measurements have been made on mice of four strains: A/Cam, CBA/FaCam, SF/Cam and Peru. Hybrid mice produced by crossing these strains have also been studied. All the stocks of mice thrive under the standard conditions in which they are kept.[18]

One of the first parameters measured on the mice of the four strains was the absolute weight of the adrenal glands. Very significant differences were found between the strains, even when allowance was made for the differences in body weight.[1, 2] Strain A mice have small adrenals, CBA mice have ones which are intermediate in weight and Peru mice have large adrenals (see Table I). Male SF mice have adrenals similar in size to those of A males, but SF females have adrenals slightly larger than those of CBA females. Thus there are strain differences in sex-difference as well as strain differences in the absolute weight of the gland. The pattern of means and variances found for the F_1, F_2 and backcross generations were those that were expected if a significant proportion of the observed differences between the parental strains was genetic in origin. The number and identity of the genes responsible for the observed strain differences are not yet known.

The mice of the four strains differ not only in the total weight of the adrenal glands but also in the volumes of the several parts of the gland.[19] The volume of the adrenal medulla,

TABLE I

CHARACTERISTICS OF YOUNG ADULT MALE MICE OF FOUR STRAINS

	A	CBA	PERU	SF	REFERENCE
Mean body weight (gm.)	24	23	13	18	
mgm. adrenal/100 gm. body weight	10·1	12·9	19·2	15·7	1, 2
Volume of adrenal cortex (mm.^{3})	0·79	1·42		0·81	18
Zona glomerulosa	+ +	+ +	—	—	19
Ratio of 11 deoxycortisol: corticosterone in vitro	0·6	1·5			3
Blood glucose (mgm./100 ml.)		167	128		16
Liver glycogen (mgm./100 mgm.)	3·2	5·5	5·8		16
Diuretic lag (mins.)		32	53		22, 24
Fraction of water-load excreted	131%	133%	100%		24, 25
Proportion of nephron occupied by convoluted tubules		62%	46%		22, 24
Resistance to spironolactone	—	—	+		22
Mean ambulation score in an open-field situation	5	20	80		

which secretes adrenaline, varies twofold amongst the strains, even after due correction has been made for differences in body weight. The X zone, absent from the adrenals of mature males, is twice as large in young adult SF females as it is in A and CBA females of the same age and body weight. The volume of the zona fasciculata, which secretes hormones like hydrocortisone, also shows very significant strain differences. CBA mice have twice the volume of fascicular tissue that A mice of the same sex and age have (Table I). Studies on F_1, F_2 and backcross generations confirmed the existence of genetic variation affecting the zone. It seems likely that only one gene locus is necessary to account for the major part of this difference between A and CBA males.[20]

In addition to these quantitative differences there are qualitative differences between the strains in the structure of the adrenal glands. The zona glomerulosa is prominent in the adrenals of mice of the A and CBA strains, but it is very poorly developed in the adrenals of SF and Peru mice.[18] Studies on CBA × Peru hybrids, and progeny tests, suggest that variation at a single gene locus is responsible for the major part of this difference between CBA and Peru. The CBA allele is dominant. The SF and Peru strains, though phenotypically similar, must be genotypically distinct to account for the appearance in the F_2 of a few mice with well-developed glomerular zones in their adrenals. This means that the physiology or the two strains must be different, even though they have adrenals which look the same.

Young adult mice differ not only in the structure of their adrenals but also in the hormones that the adrenal cortex produces. A and CBA mice have been shown to differ at a locus which affects the relative amounts of corticosteroids synthesized in vitro.[1, 3] Strain A mice produce more corticosterone (B) than 11 deoxycortisol (S) whilst CBA mice synthesise more S than B. It seems likely that in vivo the two strains of mice produce similar amounts of corticosteroids per unit body weight. This is achieved in CBA mice by the combination of large adrenals with low synthetic activity and in A mice by small adrenals which are biosynthetically very active.

There are marked differences between the strains in the development of the adrenal glands as well as between those of mature mice of the different strains. The time of X zone regression in male mice differs between the strains.[21] It has been known for some time that the age at which the X zone involutes in female mice depends on the strain of the animal.[5, 9] We have found that the X zones of the female mice of the four strains degenerate at different ages. In A females degeneration begins very early, being well advanced in eight-week-old mice.

In CBA mice it takes place later, when the mice are about twelve to sixteen weeks old. This difference between the two strains in the timing of this developmental change in an endocrine organ is controlled by genetic variation at a single locus.[22] The A allele is dominant. Degeneration of the X zone can take place in the absence of the ovaries in A mice but not in CBA mice. Thus the metabolism of sex-hormones must differ in the two strains. The pattern of inheritance of this difference between the strains is the same as that found for the difference in the timing of the process in intact mice. The genetic elements controlling the two processes might be identical.

The functions of the adrenocortical hormones are many and diverse. They include important roles in the control of carbohydrate metabolism and in the regulation of the metabolism of water and electrolytes. There are significant strain differences in the level of glucose in the blood and in the amounts of glycogen stored in the liver (Table I).[16] The kidney is an important target organ for corticosteroids. Much is known about the metabolism of water and electrolytes in the four strains of mice.[22, 24, 25] Genetic variation has been shown to affect the structure of the kidney, the handling of sodium, potassium and water during diuresis, and also the response of the kidneys to an aldosterone antagonist, spironolactone (see Table I). Some of the genes responsible for the differences between CBA and Peru mice in these characters have been identified.[24] The zona glomerulosa is generally considered to be the site of production of the hormone aldosterone, which is very important in the control of salt excretion. It is, therefore, interesting that Peru mice, which are resistant to the anti-aldosterone agent, have poorly developed zonae glomerulosae, whereas this zone is well developed in the adrenals of CBA mice, which are sensitive to this inhibitor.

Even though only four strains of mice have been studied, there is evidence for genetic variation affecting many aspects of the structure and functions of the adrenal glands of mice. It is therefore likely that comparable variation exists in human populations.

STRAIN DIFFERENCES IN THE INTERACTIONS OF THE
NERVOUS AND ENDOCRINE SYSTEMS

In addition to the differences in adrenal physiology described
above there are other differences between the strains. These
suggest that there is genetic variation which affects the interactions between the nervous and endocrine systems. Female
mice of the CBA and Peru strains were raised under crowded
conditions (twelve mice per cage) and under more normal
conditions (two mice per cage). The responses of the two
strains to crowding were different. Crowding the CBA mice
resulted in complete degeneration of their X zones by the time
they were eight weeks old. In contrast, crowding the Peru mice
resulted only in the X zones of the crowded mice being 30 per
cent smaller in volume than those of the uncrowded mice.

The ascorbic acid content of the adrenal gland is measured
in the standard bioassays for ACTH, the trophic hormone of
the zona fasciculata. Table II shows the concentrations of

TABLE II

THE CONCENTRATION OF ASCORBIC ACID (μg./100 mgm.) IN THE ADRENAL GLANDS
OF MALE MICE, MEASURED BY THE METHOD OF MINDLIN AND BUTLER.[17]

Mean of Ten Measurements, $\pm$ S.E.

	MIDDAY	MIDNIGHT
CBA	151 $\pm$ 7·9	107 $\pm$ 13·6
SF	101 $\pm$ 7·3	109 $\pm$ 10·2

ascorbic acid in the adrenals of CBA and SF males. The mid-
day values for the CBA males are significantly higher than those
for the SF males. The values at midnight are, however, the
same in the two strains. It seems possible, therefore, that the
levels of ascorbic acid change during the day in mice of the
CBA genotype but not in those of the SF genotype. These
two examples imply that the regulation of the adrenal cortex
by the central nervous system can differ significantly according
to the genetic constitution of the individual.

o

BEHAVIOUR DIFFERENCES BETWEEN THE
ENDOCRINOLOGICALLY DIFFERENT STRAINS OF MICE

Only a very short period of observation is necessary for one to be convinced that the strains differ in their general pattern of behaviour. CBA and A mice are tame, A mice exceedingly so. SF and Peru mice are both 'wild' strains in that they are active, agile and difficult to catch. Further acquaintance with these two strains, however, reveals differences between them.

One parameter of behaviour that has been measured quantitatively on CBA and Peru mice is their 'ambulation index'. This is a measure of the activity of a mouse during its first few minutes in an open-field situation. A high score is typical of an active exploratory mouse and a low score of a relatively inactive one. Peru males aged seven weeks had a mean score of 80, while CBA mice of the same age and sex had a mean score of 20. F_1 mice, bred by crossing Peru males and CBA females, resembled the Peru mice in that they were very active, with a mean score of 70. The mice of the two backcrosses (all of which had F_1 mice as mothers) were different. The backcross to Peru mice all had high scores (mean 74), whilst the scores of the backcross to CBA mice were very variable, ranging from those typical of CBA mice to those typical of Peru mice. This strain difference might prove to be due to variation at a single locus, as has the one between C3H and C57BL mice described by De Fries, Hegman and Weir.[8]

ARE THE ASSOCIATIONS OF CHARACTERISTICS
OBSERVED IN THE STRAINS COINCIDENTAL OR
SIGNIFICANT?

The four strains differ from each other in many characters, both behavioural and hormonal. Considering only a single pair of strains, CBA and Peru, one notes that there are many differences between the strains. CBA mice are larger than Peru mice, and they have smaller adrenals than Peru mice. These adrenals contain a zona glomerulosa whilst those of Peru mice do not. CBA mice are less active than Peru mice. Are some of these differences reflections of a single difference between the strains in some basic component of physiology? Or are all these differences the consequences of variation in many, independent, factors? If a pair of characters can be shown to vary independently of each other, the observed differences cannot be the result of variation in a single, basic, factor. On the other hand, if variation in one character is always accompanied by variation in the other, it is very likely that there is an underlying factor common to both characters, thus accounting directly for the association observed in the strains. An example of two characters associated in a pair of strains is provided by body weight and ambulation index. Peru mice are small and active while CBA mice are heavier and are less

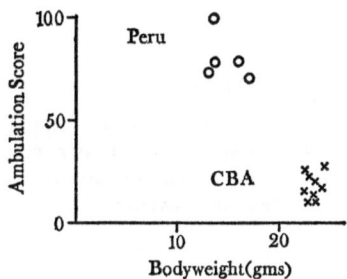

FIGURE 1

THE RELATIONSHIP BETWEEN AMBULA-
TION SCORE AND BODY WEIGHT IN MALE
MICE OF TWO STRAINS

active. Now it might be supposed that the 'activity' of a mouse is dependent on how large or how small it is. Alternatively, the size of the animal might be a consequence of how active it is. Studies of genetic segregations allow us to test these possibilities. Figure 1 shows the ambulation scores of individual mice of the two strains plotted against body weight. There appears to be an overall negative relationship between ambulation and body weight, suggesting that the difference in activity between the strains is closely related to the difference in body weight. However, when we test the association observed in the strains by crossing the strains and making measurements on mice of the generations in which segregation can take place, we find that the association largely breaks down. Figure 2 shows the observations on the backcross to CBA mice. The ambulation scores fall into two classes, one resembling each of the parental strains. The ambulation scores are no longer related

to body weight.   The major difference in ambulation between
the strains is separable from the difference in body weight
between the strains.   Thus there is no basic difference (such
as the level of thyroid activity) between the two strains which
has major effects on both activity and body weight.   There is,
however, evidence of a slight negative relationship between
body weight and activity *amongst* the highly active individuals
of the backcross, F_1 and F_2 generations.

The positive correlation between activity and adrenal
weight observed for the A, CBA and Peru strains (Table I)
does not hold for the backcrosses between CBA and Peru (see
Figure 3).   Thus in this case also, the two characters, in which

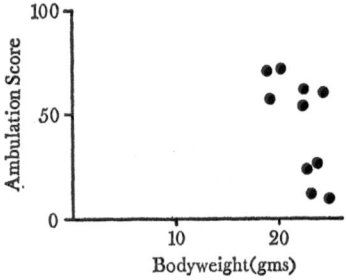

FIGURE 2

THE RELATIONSHIP BETWEEN AMBULA-
TION SCORE AND BODY WEIGHT FOR
INDIVIDUAL MICE OF THE BACKCROSS TO
CBA GENERATION

the strains differ, can vary independently and are not both
measures of a single basic difference between the strains.   Other
characters, such as the size of the zona glomerulosa, segregate
in the opposite backcross to the behaviour difference and are
thus not likely to be significantly associated with it.

The examples discussed above are cases in which correlations
observed in the strains did not hold for segregating generations.
However, cases do exist in which several characters, all showing
strain differences, always bear the same relationship to each
other, even when segregating generations are studied.   Male
mice of the CBA and the A strains differ both in the volume of
the permanent adrenal cortex and in the volume of the adrenal
medulla.   Both these parts of the adrenal gland are smaller in
A mice than they are in CBA mice.   Measurements on mice of
the reciprocal F_1s, the F_2 and both backcrosses show that there
is genetic variation affecting both characters.   Since all the

observations on the segregating generations fall on, or close to, the regression line relating medullary volume and fascicular volume in the parental strains, it seems likely that there is but one basic difference between the strains affecting these two characters.[20]

Thus by carrying out breeding experiments, such as simple backcrosses or the more sophisticated programmes described by Thoday,[26] one can further analyze characters in which significant strain differences have been found. Such approaches can be used not only to answer questions about the numbers of genetic determinants of a given set of characters, but also to

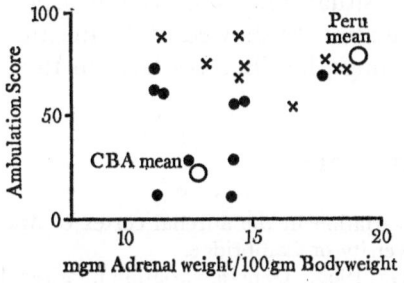

FIG. 3

THE RELATION BETWEEN ADRENAL WEIGHT (PER UNIT BODY WEIGHT) AND AMBULATION SCORE FOR INDIVIDUALS OF THE BACKCROSS TO CBA (●) AND THE BACKCROSS TO PERU (×)

The mean values of the two parental strains are also shown.

study the relationships between different characters. These characters can be measurements of endocrine activity, behavioural parameters or characters involving both these systems.

SOME CONSEQUENCES OF GENETIC VARIATION IN ENDOCRINE SYSTEMS

Certain environmental events, acting through the central nervous system, will affect the endocrine systems of every individual. The metabolic responses of mice of different genotypes, varying in the kinds of ways described in this paper, will differ even though the initial situation, and the responses of the nervous system to it, were the same in all cases. Genes which affect the endocrine systems of men and women will also result in individual people showing different metabolic responses to the same environmental situation. These different responses could result in different patterns of subsequent behaviour, differential resistance to disease, or differential reproductive success. However, in natural populations of both men and mice the problems of predicting the results of a particular environmental stimulus will be made more complex by the many genes which affect the functioning of the nervous system, in addition to those affecting the endocrine systems. Such variation has been described by Professor Vandenberg and by Professor Eysenck earlier in this symposium.[10, 27] Genetic variation affecting the individual's perception of his environment must also be considered when we wish to predict the behavioural and physiological responses of a particular person to a particular environmental situation. This is the aspect covered by the film Professor Kalmus[14] showed at the meeting and which, with his tape recordings, he discusses later in these proceedings.

REFERENCES

1. BADR, F. M. 1965. Genetic variation in the adrenal cortex of *Mus musculus*. Ph.D. Thesis, University of Cambridge.
2. BADR, F. M. and SPICKETT, S. G. 1965. Genetic variation in adrenal weight relative to bodyweight in mice. *Acta endocrinol. Copenh.*, Suppl. **100,** 92.
3. BADR, F. M. and SPICKETT, S. G. 1965. Genetic variation in the biosynthesis of corticosteroids in *Mus musculus*. *Nature, Lond.* **205,** 1088.
4. CAMPBELL, H. J. and EAYRS, J. R. 1965. Influence of hormones on the central nervous system. *Br. med. Bull.* **21,** 81.
5. CHESTER JONES, I. 1957. *The Adrenal Cortex* p. 111. Cambridge University Press.
6. CHRISTIAN, J. J. and DAVIS, D. E. 1964. Endocrines, behaviour and population. *Science* **146,** 1550.
7. COURT BROWN, W. M., JACOBS, P. and PRICE, W. H. 1968. Sex chromosome aneuploidy and criminal behaviour. In this volume (p. 180).
8. DEFRIES, J. C., HEGMAN, J. P. and WEIR, M. W. 1966. Open field behavior in mice : evidence for a major gene effect mediated by the visual system. *Science* **154,** 1577.
9. DELOST, P. and CHIRVAN-NIA, P. 1958. Différences raciales dans l'involution de la zone X. *C. r. Séanc. Soc. Biol.* **152,** 453.
10. EYSENCK, H. J. 1968. Genetics and Personality. In this volume (p. 163).

11. GRÜNEBERG, H. 1952. *The Genetics of the Mouse*, 2nd edition. The Hague. Martinus Nijhoff.
12. HASKELL, P. T. and MOORHOUSE, J. E. 1963. A blood-borne factor influencing the activity of the central nervous system in the desert locust. *Nature, Lond.* **197,** 56.
13. HUNTLEY, R. M. C. 1966. Heritability of intelligence. In *Genetic and Environmental Factors in Human Ability* (p. 201). Ed. J. E. Meade and A. S. Parkes. Edinburgh. Oliver and Boyd.
14. KALMUS, H. 1968. The worlds of the colour blind and the tune deaf. In this volume (p. 206).
15. LEVI, L. 1965. Endocrine responses to emotional stimuli. *Acta endocrinol. Copenh.*, Suppl. **100,** 19.
16. MALTBY, D. 1967. Strain comparisons of some aspects of carbohydrate metabolism. *Mouse News Lett.* **37,** 17.
17. MINDLIN, R. C. and BUTLER, A. M. 1937. Determination of ascrobic acid in plasma : a macromethod and micromethod. *J. biol. Chem.* **122,** 673.
18. SHIRE, J. G. M. and SPICKETT, S. G. 1967. Genetic variation in adrenal structure : qualitative differences in the zona glomerulosa. *J. Endocr.* **39,** 277.
19. SHIRE, J. G. M. and SPICKETT, S. G. 1968. Genetic variation in adrenal structure : strain differences in quantitative characters. *J. Endocr.* **40,** 215.
20. SHIRE, J. G. M. and SPICKETT, S. G. 1968. Genetic variation in adrenal structure : quantitative measurements on hybrid generations. (In preparation).
21. SPICKETT, S. G. and BADR, F. M. 1965. Genetic variation in the timing of X zone involution in male mice. *Acta endocrinol. Copenh.*, Suppl. **100,** 93.
22. SPICKETT, S. G., SHIRE, J. G. M. and STEWART, J. 1967. Genetic variation in adrenal and renal structure and function. *Mem. Soc. Endoc.* **15,** 271.
23. STANBURY, J. B. 1960. Familial goiter. In *The Metabolic Basis of Inherited Disease* (p. 273). Ed. J. B. Stanbury, D. S. Frederickson and J. B. Wyngaarden. New York. McGraw-Hill.
24. STEWART, J. Genetic variation in the metabolism of water and of electrolytes in mice. Ph.D. Thesis, University of Cambridge.
25. STEWART, J. and SPICKETT, S. G. 1967. Genetic variation in diuretic responses : further and correlated responses to selection. *Genet. Res.* **10,** 95.
26. THODAY, J. M. 1967. Uses of genetics in physiological studies. *Mem. Soc. Endocr.* **15,** 297.
27. VANDENBERG, S. G. 1967. Primary mental abilities or general intelligence ? Evidence from twin studies. In this volume (p. 146).
28. WILLIAMS, R. J. 1956. *Biochemical individuality*. New York. Wiley.

THE WORLDS OF THE COLOUR BLIND
AND THE TUNE DEAF

H. Kalmus

Galton Laboratory, University College, London

After the formal papers, and by way of contrast, I concluded this Symposium by showing a film demonstrating the subjective appearance of various environments to a colour blind person, and by playing some tape recordings of tests used for the detection of tune deaf people.

The defective perceptions of people suffering from these inherited disabilities may be visualized by simulating their preceptions—so far as we can know them—and exposing normal people to appropriately manipulated stimuli. Such an approach provides the best means of realizing the difficulties of colour blind and tune deaf people and of considering how they might be helped.

These two groups of deficiencies, inherited partial colour blindness and familial tune deafness, differ in many aspects. The sex-linked forms of defective colour vision are well-defined Mendelian unit characters, whereas tune deafness, though undoubtedly of familial occurrence, has a more complex mode of transmission. The forms of ' colour blindness ' with which we are concerned are stationary throughout life and incurable, whereas tune deafness is somewhat changeable, may improve with age and can often, to some extent, be overcome by training.

I showed part of an instructional film [1] made for the US Navy by the late Commander Dean Farnsworth, which concerns, for the most part, the vision of a young woman who is deuteranomolous in one eye and normal in the other.

A person of this type is, to some extent, in a position to refute those armchair philosophers who would maintain that nobody can visualize other people's perceptions and in particular that normal people cannot have the sensations of the colour blind.

To switch from the colour normal *Umwelt* to that of the colour blind our subject simply opens either the right eye or the left eye; she is thus in a position to describe, in the terms of normal colour vision, the tints seen by her defective eye and can match and help to reproduce the perceptions of a colour blind person when looking at, for instance, a colourfully furnished room, as was shown in the film. The film also showed how a spectrum, a microscopic slide of stained bacteria, a set of coloured cables or billiard balls are seen by different types of colour blind people. It appears that the most prevalent types of colour defectives— those suffering from protan or deutan defects—lack the red and green sensations to various degrees. Their extreme representatives see the world only in yellows, blues and greys.

Tune deaf people [2, 3] who were formerly called tone deaf are usually discovered at school by their singing in monotone, and by being unable to recognize well-known melodies. They cannot tune keyboard instruments properly nor play string or wind instruments. So far, no person has been found who is tune deaf in one ear only and thus one cannot be very definite in showing how they perceive music. All I could do was to play a few recordings indicating the tolerance which tune deaf people show in accepting very faulty, even hilarious, musical performances and to demonstrate some acoustic tests, mostly based on this inability; these were a sample of the records made many years ago with my colleagues Professor Penrose and Professor Fry. The test consists of twenty-five well-known song tunes rendered once correctly and once with one or two wrong notes all in random order. Correctness or incorrectness are recorded on a sheet of paper. Normal people would find the acceptance of the faulty tunes quite ludicrous and incredible; but the small percentage of tune deaf who hear them would not be particularly perturbed and, like the colour blind during the earlier showing of the film, would not quite see what my demonstration was all about.

The scores from these distorted tunes have been correlated with two of Seashore's tests for musical ability,[6] namely that for pitch discrimination and that for tonal memory as well as a test for number memory devised by ourselves; samples of all three tests were played. The results of all the tests are fairly highly correlated and it is possible to calculate various discriminant functions which will divide people into 'normal' and 'tune deaf'. Familial distribution appears then to be compatible with autosomal inheritance, either recessive or dominant; however, the reasoning in such procedures is rather forced, and definite conclusions would be premature.

Colour blindness, and possibly tune deafness, may be of some disadvantage in various environments and it is possible that their prevalence to-day is a consequence of relaxed selection.[5] However, this too is highly speculative. Similarly, the effect of these minor disabilities on mating preference is an unsolved problem. In view of the demonstrated positive assortative mating of the genetically deaf and negative assortative mating, or at least absence of assortative mating in the blind,[4] studies in this direction might be of some value.

REFERENCES

1. *Colour Vision Deficiencies.* The film, MN 8246, made by Cdr. DEAN FARNSWORTH for the Research Division of the Bureau of Medicine and Surgery. Obtainable from the Department of the US Navy, Washington 25, or from US Embassies.
2. FRY, D. 1948. An experimental study of tone deafness. *Speech.* March, 1948.
3. KALMUS, H. 1948. *Tune deafness and its inheritance. Proc. 8th Int. Cong. Genet.* p. 605.
4. KALMUS, H. 1966. Sense perception and behaviour. In *Wenner Gren Symposium on Anthropological Research 1964 Summer Session,* p. 33.
5. POST, R. H. 1962. Population differences in red and green colour vision deficiency : a review and query on selection relaxation. *Eugen. Q.* 9, 131.
6. SEASHORE, A. 1952. *Measures of Musical Ability.* Victor Gramophone Records. RCA, Camden, N.J., USA.

CONCLUSION

C. O. CARTER

MRC Clinical Genetics Research Unit, Institute of Child Health, London

THE future of preventive medicine, the prevention of behaviour disorders included, will lie more and more in recognizing those genetically at risk and protecting them from the additional environmental impulses which push them over into pathological states. The measures of protection may be so specialized in detail as to be quite inapplicable to the population as a whole.

Dr. Shire has dealt with genetically determined variation in the activity of the adrenal gland in mice. There is an excellent example of individual prophylaxis for such an abnormality in children. The female pseudo-hermaphrodites, due to congenital adrenal hyperplasia, often had, until fairly recently, to be brought up as boys. They were sometimes so virilized both in behaviour and in physique that there was no alternative. They can now be treated entirely effectively with cortisone. They can grow up as normal girls, and the first few to have been so treated are by now having children of their own.

Dr. Court Brown has discussed the problem of men with the XYY syndrome, many of whom run into serious trouble. Such individuals are, however, ascertainable at birth, and clearly we have got to start screening at birth for this condition. Then it will be a challenge to the mental health services to protect them from developing into psychopaths, and I think it is reasonable to hope that their disability will be, at least, greatly mitigated, provided their special risks are recognized.

Again, in the more ordinary field of delinquency and psychopathy, there is the important question of the very early recognition of children at risk, perhaps by the kind of special tests described by Professor Eysenck. These children, perhaps shown to be highly introverted and highly neurotic on these same tests, should be protected from the additional environmental trauma—such as separation from the mother—that finally pushes them into neurosis or psychosis.

Unless we are prepared to recognize the existence of special genetic risks, and to make special efforts to identify and protect those at risk, we shall not be able to prevent a considerable fraction of behaviour disorder.

INDEX OF SUBJECTS

INDEX OF AUTHORS